THROUGH AN UNKNOWN COUNTRY

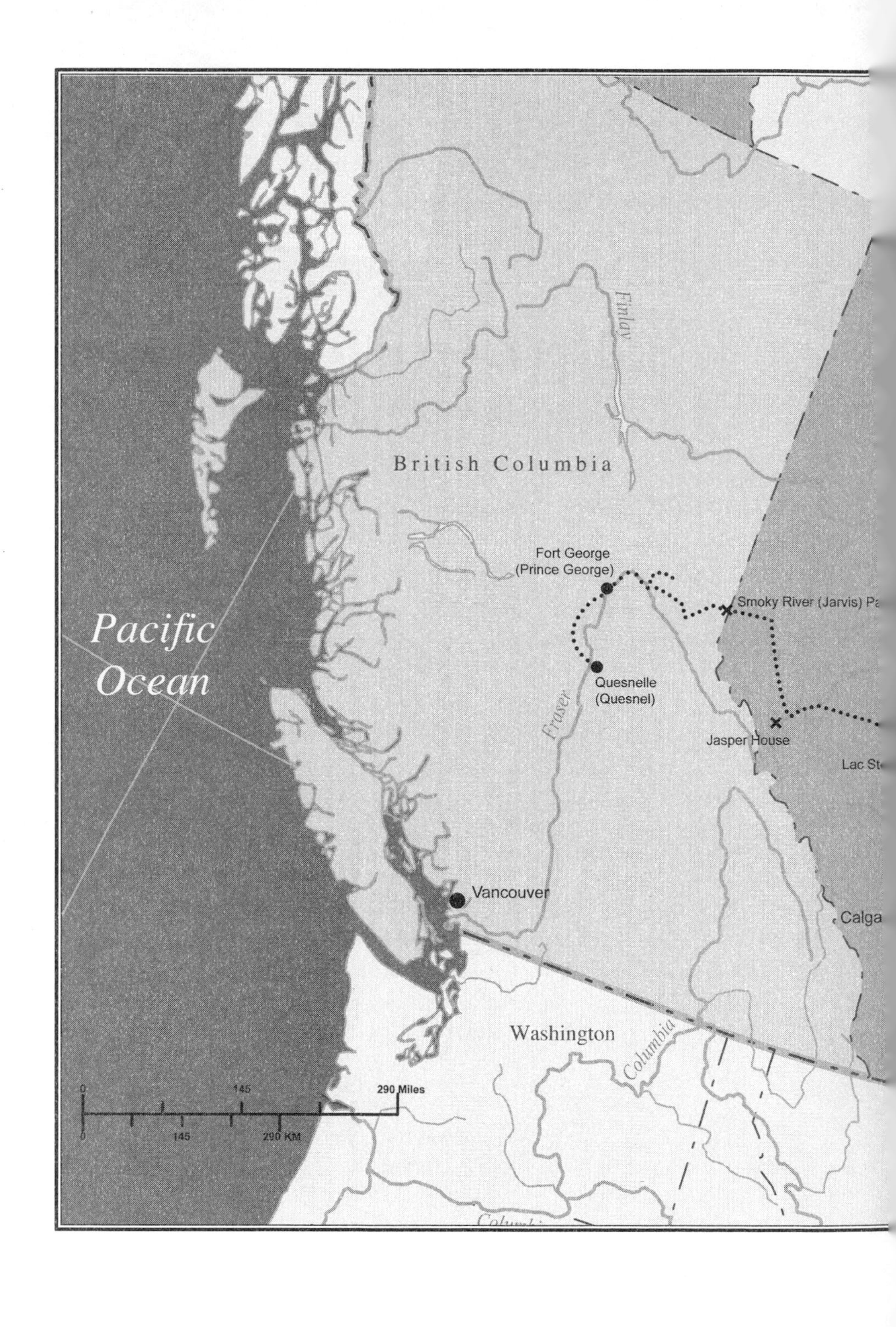

British Columbia
Finlay
Pacific
Ocean
Fort George
(Prince George)
Smoky River (Jarvis) Pa
Quesnelle
(Quesnel)
Fraser
Jasper House
Lac Ste
Vancouver
Calga
Washington
Columbia
0
145
290 Miles
0
145
290 KM

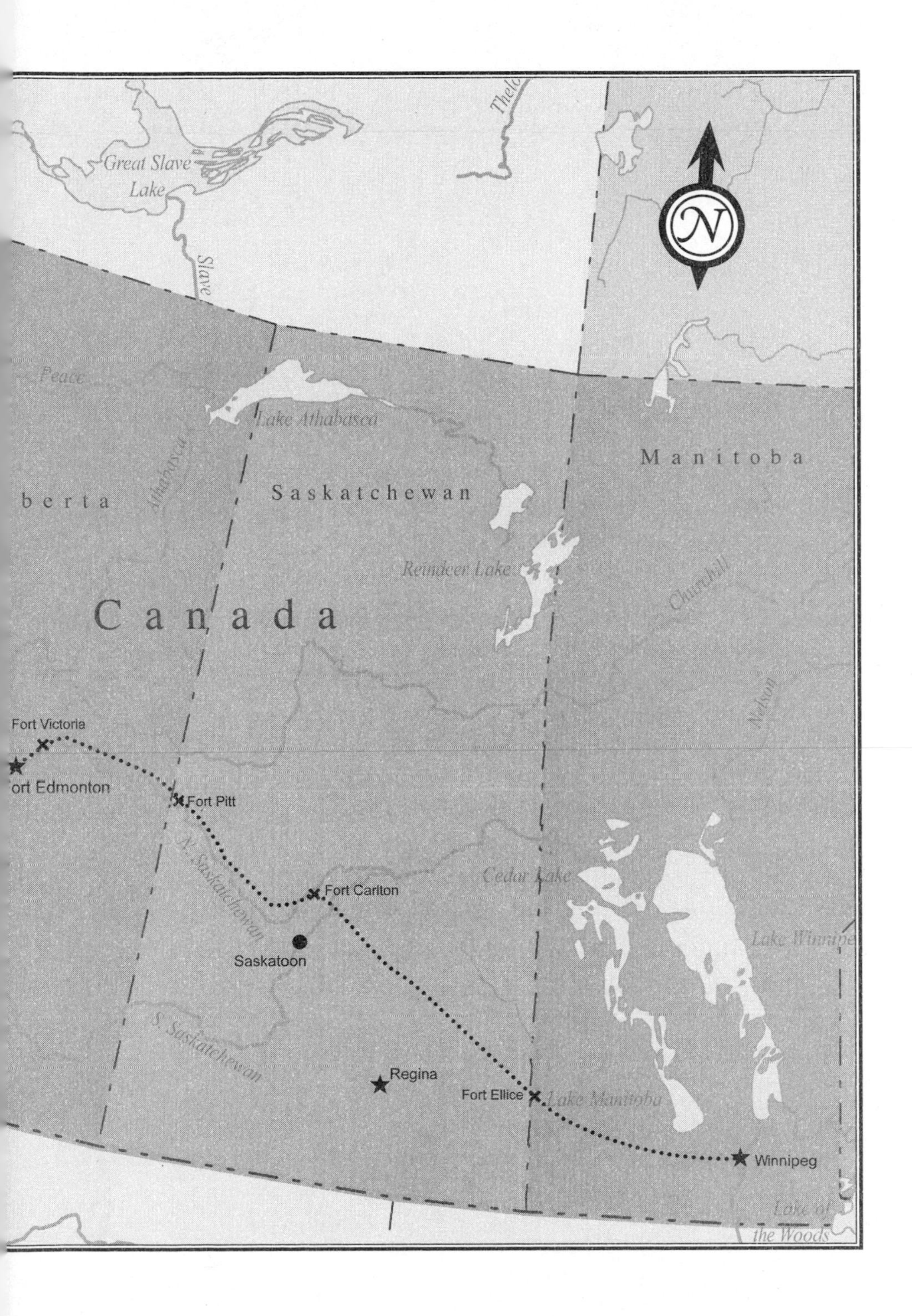

Great Slave Lake
Slave
N
Peace
Lake Athabasca
Athabasca
Manitoba
Saskatchewan
Reindeer Lake
Churchill
Canada
Nelson
Fort Victoria
Fort Pitt
N. Saskatchewan
Cedar Lake
Fort Carlton
Saskatoon
S. Saskatchewan
Regina
Fort Ellice
Lake Manitoba
Winnipeg
Lake of the Woods

THROUGH AN UNKNOWN COUNTRY

The Jarvis-Hanington Winter Expedition through the Northern Rockies, 1874–1875

Edited by

Mike Murtha and Charles Helm

Foreword by

Robert William Sandford

First edition

RMB | Rocky Mountain Books Ltd.
rmbooks.com
@rmbooks
facebook.com/rmbooks

Cataloguing data available from Library and Archives Canada

ISBN 978-1-77160-133-7 (bound)
ISBN 978-1-77160-134-4 (html)
ISBN 978-1-77160-135-1 (pdf)

Printed and bound in Canada

Distributed in Canada by Heritage Group Distribution and the U.S. by Publishers Group West

RMB | Rocky Mountain Books is dedicated to the environment and committed to reducing the destruction of old-growth forests. Our books are produced with respect for the future and consideration for the past.

We acknowledge the financial support of the Government of Canada through the Canada Book Fund and the Canada Council for the Arts, and of the province of British Columbia through the British Columbia Arts Council and the Book Publishing Tax Credit.

Nous reconnaissons l'aide financière du gouvernement du Canada par l'entremise du Fonds du livre du Canada et le Conseil des arts du Canada, et de la province de la Colombie-Britannique par le Conseil des arts de la Colombie-Britannique et le Crédit d'impôt pour l'édition de livres.

Canada Council for the Arts Conseil des Arts du Canada

"We are travelling through an unknown country without a guide and take things as they come."

—CHARLES FRANCIS HANINGTON,
JOURNAL ENTRY, FEBRUARY 16, 1875

Table of Contents

Foreword

"WE CANNOT TELL IN COLD PRECISION
WHITHER WE GO."
—JON WHYTE, "WINTER JOURNEY," 1984

Though this book is about an adventure, it is also a rare adventure in itself. Imagine an utterly classic early exploration account that has been lost in dispersed archives for nearly 150 years after being discovered. Imagine that account then being painstakingly reconstructed and lovingly brought back to light. Imagine a mystery suddenly solved; a nearly forgotten story finally told. Imagine the past suddenly reconciled with the present – in both word and image – in a manner that puts the future into relief. This is such a book.

The importance of the construction of the Canadian Pacific Railway (CPR) to the history of Canada can be gauged by the fact that 130 years after the completion of that railway important historical accounts continue to be put before a public still eager to learn more about that formative period in the life of our country. This book is exciting in that it tells the publicly untold story about a virtually unheard-of expedition sent out west to exhaust all possibility that there might be a more desirable transcontinental rail route across Canada to the Pacific than the Yellowhead Pass. The goal of the expedition was to determine the practicality of Smoky River Pass as a possible rail route. This, remarkably, is the story of two men, one from Prince Edward Island and one from New Brunswick, who, with almost no mountain experience whatsoever, crossed the Rocky Mountains on foot and by dogsled in the dark and bitter cold of winter. Such was the political urgency of finalizing a rail route across Canada that Edward Worrell Jarvis and Charles Francis (Frank) Hanington headed into what was, to them at least, the unexplored interior of northern British Columbia in early December 1874. A month later, they headed without a guide into what remain even today some of the remotest regions of the Canadian Rocky Mountains.

Jarvis and Hanington are important figures in the early settlement history of the Canadian West and significant in particular in the Rocky Mountains upon the map of which they have left their names. They were witnesses to history. They travelled across the country at a moment that at once marked the end of the fur trade that founded the country and the beginning of a new West that marked the arrival of people like Gabriel Dumont and institutions like the North West Mounted Police, the first detachment of which they met as it headed west into history. As this book will demonstrate, Jarvis and Hanington kept an astonishing record of what they saw. Because they kept a daily record of temperatures and described the nature of the rivers over which they travelled, their weather and hydrological observations are of interest to us even today. In our time, the 165-day, 3000-kilometre Jarvis and Hanington expedition of 1874–1875 would be heralded as a heroic human epic and the two would be celebrated here and abroad as great Canadian adventurers, extreme athletes, heroes. Unfortunately for Jarvis and Hanington, however, such epics were in their time a matter of historical course. What is really groundbreaking about this book is that it puts this epic into historical perspective. It brings together all the hitherto uncollected versions of the same expedition account and does so with the accompaniment of both historical and modern photographs of the places along the nation-spanning route travelled by the expedition. In a way, we now know where they were better than they did, and as a result we know why the expedition was so important.

It is very interesting to compare the documents in this book. The official report of the expedition contains the usual silences and understatements concerning hardship and suffering. The report does not indicate how early the expedition was in the settlement history of the West. It is silent on how completely on their own they were without a guide. That the expedition was travelling light is an understatement. They had no tents. Each man carried a cotton sheet out of which each could fashion a *tente d'abri* as Sandford Fleming described it. The report did not dwell directly on the fact that they became desperately lost. It does not mention that the expedition faced snow depths and extremes of cold temperatures that are uncommon today. It does not mention that for all the help they would have got, if they could not have carried on during the bitter January of 1875 they might as well have been on the dark side of the moon. (The report does note, however, that the ability to suffer without complaint was described as "pluck," something men at that time were expected to possess in addition to a stiff upper lip if they were to be selected for such expeditions.)

The expedition was in such serious condition by the time it had conducted its official exploration duties that it had to abandon nearly everything but food to make a desperate forced march over rugged terrain to the Athabasca Valley with the hope of making it to the Hudson's Bay trading post at Jasper House. The men nearly starved and many of their dogs died in the deadly dash. The post, however, was abandoned and the party did what every expedition did when they were in trouble: they relied on the humanitarian sensibilities and help of local Aboriginals.

It is Frank Hanington's letters that reveal the most about the expedition and the West through which they passed during a crucial turning point in its history. These letters illustrate that Hanington was not only a keen observer but a good writer and a diligent chronicler of what he saw and experienced. Though Frank Hanington's letters to his brother Edward back in New Brunswick do not account for every day, they account for every step of the journey, which Hanington appears to have counted. (He notes that it is 2,188,900 paces from Quesnelle to Edmonton). Hanington's later reminiscences are particularly revealing in that he had clearly had time to reflect on the expedition and to give added meaning and value to his experiences.

Though the Hanington letters are longer and more descriptive, Jarvis's diary entries are also important. As they place Jarvis and Hanington within the context not just of their time but within the larger circumstance of family and place, the illustrated biographies provided by Murtha and Helm are particularly valuable. As if a spell had been cast over them, both Jarvis and Hanington returned west after the expedition. It is interesting to note that Jarvis later joined and became a major figure in the North West Mounted Police, whose westward march intersected with his own journey home. Frank Hanington returned to British Columbia as an engineer working on the completion of the main line of the Canadian Pacific Railway through the Thompson and Fraser river canyons and later on the rail line through Crowsnest Pass.

No foreword claiming to represent a larger perspective on the value of this book would be complete if it did not acknowledge the hard work of Mike Murtha and Charles Helm in bringing this important account back from the brink of being lost to public memory. Because of their persistence and excellent scholarship, readers can, for the first time, examine all of the expedition narratives together. Highly relevant photographic images bring these accounts to life. Containing as they do Jon Whyte's epic 1984 poem on the expedition entitled "Winter Journey," even the appendices are of considerable value, making every page in this volume worthwhile.

As Murtha and Helm note in their introduction, the Jarvis-Hanington story really is an exploration classic; an account that deserves to live on in Canadian history. This book is important enough historically and well enough written and edited to prove Frank Hanington may have ultimately been wrong when he said near the end of his life that Canadians don't give a damn about their own history. As this book attests, Canadians like Mike Murtha and Charles Helm certainly care, and now because of this book, so will many, many others.

Robert William Sandford
Canmore, Alberta

Acknowledgements

The editors would like to thank the following individuals and institutions who assisted with the research and provision of material for this book: Robert Allen, Association of BC Land Surveyors; Reg Arbuckle, Alberta Parks; Jean Guy Bergeron; Brian Carnell; Wayne Giles; Marney Gilroy, Cumberland County Genealogical Society, Nova Scotia; Gordon Goldsborough, Manitoba Historical Society; Lena Goon, Whyte Museum and Archives of the Canadian Rockies; Chris Goulet; Chris Kotecki and Sharon Foley, Provincial Archives of Manitoba; Frances Macdonald; Mary Macdonald; David MacLennan (C.F. Hanington's great-grandson) and other members of the Hanington family; Leslie Middleton, Quesnel and District Museum and Archives; Keith Monroe; Mike Nash; Bill and Bev Ramey; Graham Reynolds; Nick Richbell, CP Archives; Rick Roos; Robert Sandford; Jay Sherwood; Kevin Wagner, BC Parks; Paul Walsh; Glenbow Museum and Archives; Kamloops Museum and Archives; National Gallery of Canada, and the Vancouver Public Library, local history section.

We are especially grateful to the Provincial Archives of Manitoba for permission to publish excerpts and sketches from Jarvis's diaries and to the BC Archives for permission to publish excerpts from Hanington's reminiscences. None of this material has been previously published. Library and Archives Canada kindly authorized the republishing of Jarvis's report and Hanington's journal.

We would also like to thank the following for permission to use material: Boone and Crockett Club (*From the Peace to the Fraser*); Whyte Museum of the Canadian Rockies (Fay excerpts); Mary Andrews (Andrews excerpt); Harold Whyte (Jon Whyte's poem); McFarland & Company, Inc. (Putnam excerpt); and the Alpine Club of Canada and Lawrence White (excerpts from the *Canadian Alpine Journal*). Excerpts from *The National Dream: The Great Railway, 1871–1881* by Pierre Berton are reprinted by permission of Anchor Canada/Doubleday Canada, a division of Random House of Canada Limited, a Penguin Random House Company.

We are indebted to Don Gorman and Rocky Mountain Books, who have been enthusiastically supportive of this project from its inception. And our thanks to Kirsten Craven for her careful edit.

Finally, and not least, we wish to thank our wives Patricia and Linda, who tolerated the time we spent on the research and preparation of this book and who have been supportive throughout.

Introduction

In the winter of 1874–1875, Edward Worrell Jarvis and Charles Francis Hanington undertook a dangerous and challenging expedition on behalf of the Canadian Pacific Railway Survey from Quesnel in British Columbia to Winnipeg, Manitoba. In the depths of a bitterly cold winter, it led them over the northern Rocky Mountains, searching for Smoky River Pass. The trip took 165 days and covered over 3000 kilometres, of which almost 1500 kilometres were on snowshoes. They were exploring unknown terrain, they lost most of their sled dogs to exhaustion and starvation and they themselves almost starved when their food ran out. But they doggedly struggled on and they and their Aboriginal companions survived to tell a remarkable tale. It's a tale worth retelling.

In 1871, British Columbia joined the four-year-old Canadian federation. A commitment of the federal government was the construction of a railway connecting British Columbia to the rest of Canada within ten years. A similar commitment, the Intercolonial Railway, had previously accompanied Nova Scotia and New Brunswick's entry into the federation and was under construction. The railway to British Columbia would complete the "National Dream," a railway between the Atlantic and Pacific coasts to unite the enormous new country and fend off American interest in drawing parts of the new Canada into the United States. America had twice invaded Canada, and it was essential to unite the former British North American colonies to withstand the continuing threat. "Manifest Destiny," the US presumption of a right to the whole North American continent, was alive and well. It was only a few years since the United States purchased Alaska, and British Columbia stood in the way of a connection to the rest of the country.

The Canadian Pacific Survey was appointed in 1871, under the leadership of Sandford Fleming, to start work immediately on the proposed railway, the Canadian Pacific. This was Prime Minister John A. Macdonald's vision of the catalyst for a great transcontinental nation, great both in physical size and in stature. In other parts of the world nations might build railways – here a railway would cement a nation. It was bold and visionary: the longest railway ever envisaged, through uncharted, difficult terrain. Its realization was an enormous logistical and financial challenge for the small population. The railway would have to pay for itself by opening the prairies to agriculture and settlement and by accessing the timber, mineral and tourism resources of British Columbia.

First a route had to be determined, a practical, efficient and cost-effective route. The sheer size and scale of the project was an issue of its own. And geography was simply not on Fleming's side. Starting in the east, there was the barren granite of the Canadian Shield, followed by many miles of seemingly bottomless muskeg. Where the muskeg gave way to the prairies, the going was easier, but there was no clear indication of where the route would or should go, and there was a shortage of suitable timber. Then came the big barrier – the Rockies – punctuated by passes of varying height and latitude. No one really knew which of these would be the most suitable, partly because once through the Rockies there was still more trouble. Depending on where these mountains were breached, there would be other mountain ranges with even worse weather and maybe fewer passes. There followed an array of nightmarish canyons to descend to penetrate the Coast Mountains, followed by a choice of harbours.

Because of these geographic influences, the number of options, and hence the number of necessary surveys and the amount of required exploration, increased with distance west from Lake Superior. In time, Fleming would be criticized for being too meticulous, sending out too many surveys for too long and at too great an expense. Yet it is hard not to sympathize with his plight – knowledge of the interior of British Columbia was rudimentary at best, and there was always the chance of a new and better pass or route.

As the surveying progressed, various passes through the Rockies were eliminated from consideration and the Yellowhead Pass became Fleming's preferred option and was endorsed by the government in April 1872. But in January 1874, his chief engineer in British Columbia, Marcus Smith, passed along a report of another pass a little to the north that, if feasible, would provide a direct and much shorter line between Edmonton and Fort George (now Prince George), if a northern route to tidewater were chosen. This Smoky River Pass had to be investigated – and time was running out. The railway to British Columbia's coast was supposed to be built by 1881 and yet route exploration was still underway. It was already 1874 and so the solution was a winter survey – a dangerous undertaking through avalanche terrain in a heavy snowbelt at the coldest time of the year.

Fortunately, Fleming had two ideal candidates available and already working in the area, two civil engineers still in their twenties – Edward Worrell Jarvis of Charlottetown, Prince Edward Island, and Charles Francis Hanington of Shediac, New Brunswick. Jarvis had worked for Fleming on the Intercolonial Railway from 1867 to 1871. Both joined the Canadian Pacific Survey in 1871 and worked together for the next four seasons on preliminary route-finding surveys. During the summer of 1874, they had completed an exploration of the North Thompson and Clearwater valleys, north of Kamloops, followed by a survey down the upper Fraser River from Yellowhead Pass. They were highly regarded. In 1873, Smith had written to Fleming:

> For the survey between the mouth of the Clearwater and Howe Sound, a distance of about 280 miles, nearly half of which is through the rugged Cascade Mountains, you could only allow me two parties, with which there was barely time to complete the work before the close of the season, as it was now the 1st of June. However, you gave me two well organized parties, viz., Divisions M and X under Mr. E.W. Jarvis and Mr. C.H. Gamsby respectively.[1]

High praise indeed. Fleming had his men and they were both willing to accept the challenge.

After completing their summer's work, Hanington cached some supplies along their proposed route up the McGregor River valley (then known as the North Fork of the Fraser River). Then they travelled to Quesnel to acquire additional supplies and prepare for the long expedition. At that time there was only a small Hudson's Bay Company store at Fort (Prince) George. Quesnel, by contrast, was a booming town and supply centre for the Cariboo gold fields. This was their starting point.

They left Quesnel on December 9, 1874, and finally arrived in Winnipeg on May 21, 1875.

The only guidance they had was Fleming's 1874 map showing the routes that had so far been explored and an "Unexplored Route" following the North Branch of the North Fork of the Fraser River through Smoky River Pass at approximately 54° 30'N (see Figures 1 and 2). Jarvis and Hanington's challenge was to follow this route, find the pass and change the description to "Explored." And this they did.

After the completion of the expedition, Jarvis wrote an official report and narrative for Fleming, eliminating the Smoky River Pass from further consideration because of its altitude and steep gradients. Hanington prepared a map of their route. The report was published in Fleming's 1877 progress report. And in early 1876, Hanington described the expedition in a series of letters to his brother that were eventually deposited at the National Archives and published in 1888. Both accounts were published in obscure government reports and not for a general audience, and yet together they provide vivid descriptions of an outstanding trip. They are brought together here for the first time. Additionally, in preparing this book we have discovered Jarvis's diaries, containing previously unknown sketches, and Hanington's reminiscences of his career. Neither is known to the general public. They supplement the previously published material, and relevant excerpts are included. Jarvis's sketches are particularly significant as the expedition could not carry heavy and fragile photographic equipment and so there is no photo record.

There are three compelling quotes that, in our view, encourage the editing and publication of these accounts. The first is by Gerry Andrews, who from 1923 to 1925 was the first schoolmaster of the small Métis and Cree settlement of Kelly Lake, just west of the British Columbia – Alberta border in the Peace Region. In 1925, Andrews undertook a packing trip south to Jasper, staying east of the Rockies and crossing the route of Jarvis and Hanington. In his book, *Metis Outpost*, Andrews pays homage to that epic adventure, and concludes with these lines: "Jarvis's vivid narrative of that epic journey...should be prescribed reading for all Canadian high schools, along with better-known classics such as the journals of Alexander Mackenzie, Simon Fraser, David Thompson and Paul Kane."[2]

The second quote is from C.F. Hanington himself, in a March 1926 letter to Samuel Prescott Fay, who in 1914 had led a party of five men and 21 horses from Jasper to Hudson's Hope along the eastern flanks of the Rockies. His route ran partly parallel and partly perpendicular to that of Jarvis and Hanington and overlapped in one short stretch: Jarvis Lakes in Jarvis Pass. More than half a century had elapsed since Hanington had visited and described that pass. Fay managed to contact Hanington, offering copies of his photographs, to which he received the following response:

> My dear Sir
>
> I am infinitely obliged for yours of 6th inst, and surprised that you should know anything of the trip you speak of. I should very much like to have a copy of your explorations, and it amuses me to think that after 50 years I should have a chance to see pictures of that part of the country…
>
> I won't bother you with a lot of stuff, which might interest you perhaps, or it might not; but it is of little value to modern people, who don't give a Damn for what has happened in the past.
>
> Again thanking you for your letter, I am
>
> Yours faithfully
>
> C F HANINGTON
> Major C.F. Hanington C.E. M.E.I.C.[3]

The third quote is from archivist Douglas Brymner, who published Hanington's journal. In his 1887 report to Parliament, he observed that

> [t]he narrative by Mr. Jarvis is very interesting, but it, of necessity, fails to give those minute details and personal feelings which are to be found in Mr. Hanington's journal, the one being a narrative drawn up for publication, with all the reserve which that fact implies, and the other written from day to day unreservedly and whilst every impression was fresh and the most trifling incidents fully remembered. Both narratives should be read together.[4]

These words, we believe, are as true today as they were in 1887. So we thank Mr. Brymner for his suggestion, and comment that a century and a quarter later it is not too late to make it a reality. However, we take issue with Major Hanington, and this book is in a sense a rebuttal of his fatalistic dismissal of "modern people." We do indeed "give a Damn for what has happened in the past" and for what he and Jarvis achieved, and we are pleased to celebrate it. We hope, also, that this book can be a catalyst for the vision of Mr. Andrews so that the gripping details of a forgotten winter in the northern mountains and the prairies take their place alongside other classics of Canadian exploration.

The reports and journals are concise and well written and are characterized by understatement rather than hyperbole. They transport us to a frozen world of snowshoes and dog teams, of sextants and compasses and moccasins, a world devoid of the ability to call for help when needed. In this world, Jarvis and Hanington wrote in reserved terms of the triumph of their spirit over privation and adversity in unbelievably trying circumstances. In a world used to high-tech survival gear, GPS devices and satellite phones, their story is sobering and inspiring. We invite you to step into this world and, in armchair fashion, engage with the snows and tribulations of the winter of 1874–1875.

CHAPTER 1

E.W. Jarvis's Report and Narrative

REPORT ON EXPLORATION ACROSS THE ROCKY MOUNTAINS BY SMOKY RIVER PASS[1] BY E.W. JARVIS. WINNIPEG, MANITOBA, 24TH MAY, 1875.

SIR, – Having received your instructions by letter and telegraph, with reference to the exploration of a pass said to exist through the Rocky Mountains at the head waters of the Smoky River[2] (an important branch of the Peace River, and so named from the smoke arising from burning seams of coal near its mouth),[3] I made the necessary preparations for an extended winter trip, and left Quesnelle[4] mouth, B.C., on the 9th December last, with one assistant, one dog driver, and an Indian boy as cook. As the River Fraser was still open, I followed the "Telegraph" trail[5] to the crossing of Blackwater River,[6] and thence the trail opened by Mr. Bell,[7] last summer, to Fort George,[8] where I procured dogs from the Hudson's Bay people. After some delay in obtaining Indians to accompany me, and endeavouring (unsuccessfully) to procure a guide, I finally left Fort George on the 14th January, the ice on the Fraser River having only set fast during the intense cold of the preceding week. The party now consisted of eight men and six dog-trains, carrying provisions calculated to last two months.

A few miles above the Giscome Portage[9] I left the main river, and, following the "North Fork,"[10] kept as nearly as possible to the line marked "Unexplored Route" on sheet No. 8 of the plans accompanying your report of January, 1874.[11] Where the stream again divides I took the left hand, or North Branch[12] (as it appeared to offer greater facilities for a line), and followed it to its source, a semi-circular basin in the heart of the Rocky Mountains, completely closed in by glaciers and high bare peaks. As there was evidently no pass in this direction, I returned 60 miles to the Forks, and decided to try the South Branch.[13] From this point, in the middle of February, I sent two of my Indians with two of the hired trains to Fort George, and wrote to you by that opportunity. The party, now reduced to a minimum, was put on regular allowance of provisions, and I enjoined on everyone the absolute necessity of strict economy both of supplies and time.

For the first 48 miles from the "Forks" the valley of the South Branch was very favourable to an easy line, and though at that point it turned sharply to the north-east and entered the main range of mountains, the stream still continued to rise with very easy grades. The valley here was half a mile wide, with the hills rising at a slope of 1 to 1 on each side, and thinly timbered with spruce, black pine and a few poplars. At 71 miles from the "Forks" (or 169 from Fort George), the mountains close in, and in a distance of eight miles the river rose from an altitude of 3,200 feet above the sea to 5,300 feet, at which elevation, four miles further on, I reached the summit lake[14] and crossed the divide on the 25th February.

Although the altitude of the pass at once showed it impracticable for railway purposes, I decided that, having gone so far, I would push on to the River Athabasca,[15] in order to obtain some topographical knowledge of the country lying between it and the mountains; and, accordingly followed the river running east from the divide for 87 miles, which I believe to be the head of one branch of the Smoky River.[16] Here it turned away to the northeast, and I considered it useless to follow it any further, my object now being to reach the River Athabasca as soon as possible, and seek relief at one of the Hudson's Bay Company's posts on that river.

Having lost several of my dogs (from frostbite and exhaustion), I was obliged to make a cache of my instruments, &c., and leaving everything not absolutely necessary at the cache, we started overland in a south-easterly direction, each carrying his blankets and his share of the provisions. After 108 miles of very difficult travel over a terribly broken country, crossing high parallel ridges and the intervening valleys (in all of which the water runs north-east, or in a similar course to the Smoky River and the Athabasca), and which occupied no less than 11 days, I reached the "Fiddle River" depot,[17] built by Mr. Moberly,[18] intending to obtain at Jasper House[19] a fresh supply of provisions, now nearly exhausted, to carry us to Edmonton.[20] The Company's post was, however, abandoned; but I was fortunate enough to fall in with some Indians in the neighbourhood, from whom I procured sufficient to last for six days (at one pound per man per day), but which I economized for ten days; and leaving the remnant of my half-starved dogs in care of the Indians, we shouldered our packs and marched to Lake St. Anne,[21] where we arrived on the 30th March, having lived the last three days on the anticipation of a meal at the journey's end.

Between Fiddle River and Lake St. Anne I followed a line of country some miles to the north of the line run by Mr. Moberly,[22] and north of the old H.B. trail. A good location can be got here. I was unable to explore a line from Root River to White Earth (old) Fort,[23] as you directed, owing to extreme bodily exhaustion consequent upon the hardships we underwent.

From Lake St. Anne I drove to Edmonton, and as the winter season was so near its close, I decided to proceed here as rapidly as possible, so as to be in time for the summer's work. I travelled from Edmonton to Victoria[24] with flat sleds and horses; and to Fort Pitt[25] I packed the animals, as the snow already showed signs of going. From Fort Pitt I took carts to Carlton,[26] but my progress was much impeded by the return of winter with freshly fallen snow. At Fort Carlton I remained four days, to rest the horses and wait for the appearance of some bare ground, and leaving there on the 5th inst., reached Winnipeg on the 21st.

The accompanying plan [not reproduced in Appendix H] shows the route followed, and may be taken as tolerably correct, the distances all being paced from Fort George to Lake St. Anne. Observations for latitude were taken with the sextant and with boiling-point thermometer for altitude, as far as the cache on Smoky River; but since then the aneroid and compass alone were used. A register of the minimum temperature was also kept, with notes as to the depth of snow, copies of which are subjoined.

The extreme depth and softness of the snow, together with the many heavy storms experienced, prevented any great progress being made; on several days, after working hard from daylight till dark, we could not accomplish ten miles. An abstract is given of the distances travelled and the time occupied; over 900 miles being on snow-shoes, for the last 300 of which each carried his pack. My Indians at times became much disheartened, but behaved well throughout.

We fortunately escaped accident or sickness of any kind, except the unfailing attendants on hard travel in winter – snow-blindness and "*mal de raquette*;"[27] and I am glad to be able to report the successful completion of the most hazardous expedition I have ever taken part in.

In conclusion, I must mention the generous hospitality and ready assistance I invariably received from the officers of the Hudson's Bay Company, to many of whom I was personally unknown; and I must give a word of praise to the pluck and endurance of my assistant, C.F. Hanington. A statement of the expenditure incurred will be sent you in a few days, together with such vouchers as I have been able to obtain.

I have further in preparation an account of the exploration written in more extended narrative form which will shortly be forwarded to you.

I am, Sir,
Your obedient servant,
E.W. JARVIS,
Engineer in charge of expedition.

SANDFORD FLEMING, Esq.,
Chief Engineer.

NARRATIVE OF THE EXPLORATION FROM FORT GEORGE, ACROSS THE ROCKY MOUNTAINS, BY SMOKY RIVER PASS TO MANITOBA, REFERRED TO IN THE PRECEDING REPORT.[28]

Sandford Fleming introduced the narrative in these words:

In the autumn of 1874, Mr. Jarvis was selected to make a winter exploration of the Smoky River Pass, with Mr. C.F. Hanington as assistant, and Alec MacDonald, who was engaged to take charge of the dog-trains. As this was the only means of conveying supplies it was necessary to limit the number of the party, and also to dispense with all unnecessary impedimenta.

The outfit therefore consisted of a pair of snow-shoes, a pair of blankets, and some spare moccasins for each man; while a piece of light cotton sheeting was taken to make a tente d'abri, the ordinary canvas tent being too cumbersome.

The supplies consisted of dried salmon for the dogs, and bacon, beans, flour and tea for the men; and were calculated to last two months.

In December the party pushed forward to Fort George, and there procured four dog trains with four Indian drivers, making a total strength (including those brought from Quesnelle) of twenty-five dogs and eight men. At the beginning of January the party awaited the freezing over of the Fraser, and Alec's return from Quesnelle (where he had been sent for more supplies).

At this point the narrative begins:

The weather at the beginning of January set in very cold, and we redoubled our exertions to get everything ready for a final start; the snow-shoes and sleds were being made in the Fort, but with the usual dilatoriness of Indians, who could not be made to understand the fact of any one being "in a hurry," they were already beyond the promised time of completion. Dense masses of vapor covered the river every morning, and we were pleased to see the ice stretching out from the shore on each side, and gradually setting fast all over. The thermometer all this week was down among the forties, and one morning at 6 a.m., it marked fifty three degrees below zero. Alec's return was now anxiously expected, as he was overdue some three or four days; and knowing him to be punctual we could not but fear some accident had befallen him. Every time the dogs barked some one would run to the door, being sure the wanderer had returned; but only to be as often disappointed. One morning about dawn, a dog scratched at the door – surely Alec is coming! – But it turned out to be only poor "Jack" with one leg frozen stiff. The dog, as we learned from the Indian who liberated him, had strayed back on the trail we came by and was caught in a steel trap, where he must have spent a week of intense suffering. We bathed the foot in ice and water and succeeded in getting the frost out of it, but a short time after, it mortified and the lower joint had to be amputated; the dog soon recovered, and though useless as a train dog, Mr. B.[29] kindly promised to give him shelter for the winter. I became very anxious about Alec, as well as chafing at the delay which was losing the best part of the winter; and an Indian was despatched to follow the river to Quesnelle and to return by the trail, on one of which roads he was sure to find some traces of the missing party. But the following evening, just at dark, as we were sitting round the fire, a ghost-like figure suddenly and silently appeared among us; and, the first amazement over, we gladly welcomed Alec back to the land of the living. The poor fellow was covered with ice from head to foot, and had a most spectral appearance. As soon as he was thawed out – thermometer forty-nine below zero – he told the following story: They made a good run to Quesnelle, arriving there Christmas Day, and started two days after to return by the river. There was, however, no appearance of ice at Quesnelle, so they put the dogs and load in a canoe, and taking another Indian to return with the canoe, poled their way slowly up to the Cottonwood canyon.[30] Above this there was every appearance of good going; the river was frozen over, and the canoe accordingly sent back to Quesnelle. But after a few miles up the river, the ice was found to be overflowed, and had to be abandoned for the woods on the bank; and then their troubles began. At most five miles a day could be made through the dense underbrush, and even then the process of "doubling up"[31] had to be adopted. Occasionally they would try the river again, but as this generally resulted in one or both getting an involuntary bath, it had to be given up. Their provisions also ran short; but, falling in with Indians at the mouth of Blackwater River, Alec obtained some salmon, and the services of one of the men to help him along. And finally, knowing how anxious I would be, he started ahead of the train from the Fort George canyon[32] and arrived in the manner above related. The following day the Indians and dogs arrived; and piling everything on the sleds they could possibly hold, we made all final preparations for our long tramp. The dogs from Stuart's Lake[33] had not yet turned up; but as they could not be more than four or five days distant, it was decided to push on, and wait for them at the cache made in the fall. The Hudson's Bay Co's accounts being put into shape and certified for payment, we bade adieu to our kind host, and turned our faces northward.

The party now consisted of three white men, three Indians, and three trains of dogs, and the order of march was as follows: two men in front to "break track" or beat down the snow with their snowshoes to make a road over which the dogs could travel; then the three trains, with a man driving each – the lightest being placed first – and, lastly, Hanington or myself alternately bringing up the rear and making what is called a "track survey" of the route travelled. The bearings were taken with a pocket compass, and the distance measured by pacing, forty paces to the chain[34] being found a good average on level ground or ice, and this was continued the whole of the distance to Lake St. Anne, fifty miles above Fort Edmonton. The intense cold continued until the third week in January, and camping out under these circumstances had its drawbacks. Many were the frozen noses and ears during the day's march, but, then, the exercise helped to keep us warm; while in the camp at night the largest and most roaring of fires scarcely did more than burn the side turned towards it, the other being made thus more susceptible to the cold. One curious effect of the extreme lowness of the temperature was to cause the fire to steam rather than smoke, and this with the very driest wood that could be found. The cold was also not without its effect on our four-footed companions;

they frequently had frozen toes, and we were obliged to make moccasins of flannel and leather to protect their feet. One old dog, the leader of the Cariboo train, suffered a great deal from frost-bite, and on the third day out he was noticed to be very lame all the morning. A halt was called at noon to drink a cup of tea, and "Marquis" lay down with the rest, but when a start was made the poor dog made a feeble effort to rise, gave one spasmodic wag of his tail and rolled over dead. His legs were frozen stiff to the shoulder; the minimum thermometer, exposed to the sun on top of the sled at the same time, registered forty-six below zero. A hole in the snow on the bank was the only grave we could make for him, and a spare dog being harnessed in his place the expedition pushed on, not without sincere regret at our loss. The travelling was good on the main river, there being only four or five inches of snow on the new and smooth ice, and we made pretty long marches, but the snow began to get deeper when we turned up the north branch of the Fraser,[35] at the head of which it was hoped that a pass through the mountains would be found. Six days after leaving Fort George, the cache was reached, and found not to have been disturbed either by Indians or wild beasts. Here the sleds were unloaded, and Hanington, with two trains and two Indians, went back to the main river, thence to go to Bear River,[36] where one of the Indians had a salmon cache, and to bring two sled loads of dried fish – about six hundred – back with him. Alec and I, with the other train, went forward up the north branch to explore and break track; and six days later the whole party reunited at the cache, less the one Fort George Indian, who, having handed over the salmon, had returned overland to his village. A heavy fall of snow during two days rendered the return journey tedious, and Hanington had to adopt the old plan of doubling up. The snow falls soft and moist here, and has a wonderful faculty of adhering to the sleds, as well as piling up under the bows, making killing work for the dogs.

The following day the trains, three in number, furnished by the Hudson's Bay Company, made their appearance with good loads of fish, and a most acceptable package of moccasins. The full strength of the party was now six trains (or twenty-four dogs) and eight men, part of whom only were intended to go all the way, there not being enough supplies to take them all through; some were, therefore, to be sent back when the summit was reached. A whole day was spent in loading up and "lashing" the sleds, repairing harness, moccasins, &c., and it was dark before all was ready. A small amount of provisions for dogs and men was left at the cache for those who were to return this way to Fort George. From our camp, which was on an island at the foot of the first canyon, we distinctly heard the sound of chopping on the opposite bank just as we were turning in, but no one could be persuaded away from the warm camp to solve any such mystery as this; although every one agreed there was something strange about it, no tracks having been seen; and if it were Indians, they would have been round our fire ere this. Yet there were the distinct and separate blows of the axe, and the crash of the falling tree on the river bank not two hundred yards from us, and the most careful search the following morning failed to show that any such thing had taken place. So much for the power of imagination. The great cold of last week had abated since the snow storm, and we managed to keep very snug in camp and warm at night by sleeping two together, and pretty close at that.

An early start was ordered and an early start was made, for we all saw clearly that no time must be wasted if we were to get over an unknown and apparently an unlimited distance, on a known and very limited supply of provisions. All my work of road making was useless, the heavy snow having sunk the ice and covered it with slush, completely obliterating the old track. We had to put four men ahead to make the road, the other four driving the six trains, and even then the progress was very slow. Just before noon one day Alec went to the bank where a small rill was dripping over a rock, for a drink, when he suddenly disappeared, the ice having given way and the water being deep at the foot of the rock. Johnny, however, was quick enough to make a grab at his head as he re-appeared, and beyond the wetting no damage was done. In the course of the few succeeding days nearly every one had a similar experience, though not such a complete ducking; on one occasion I went through only as far as the waist, catching the ice on both sides with my hands, but the current caught the snow shoes, and, turning them upside down, held them as in a vice, and the united efforts of all were required to extricate me. Hanington, being longer of limb, generally escaped by throwing himself flat on his face, when his body would land sufficiently far from the hole to be on sound ice; but the ice soon set firmer, and unless in the vicinity of open water was always safe.

The valley of the river we were following was about a mile wide, and running directly south-east as far as the eye could reach; on both sides were high rocky peaks covered with perpetual snow, those on the right bank being spurs of the main chain of the Rocky Mountains along whose base we were travelling. There were apparently no obstructions to an easy passage within sight, but we were sadly deceived, for less than fifty miles from the cache we found ourselves at the bottom of a one hundred foot fall,[37] with thickly timbered hills six hundred feet high on each side of it: these rising abruptly from the water's edge seemed to offer no footing for a snow shoe, much less a practicable trail for a dog-sled; but after half a day's careful exploration the only practicable plan was adopted and a regular track graded round the face of the bluffs. The great depth of the snow was serviceable to us here, for with snow shoes as shovels, and poles and brush to make bridges across the intervening gullies, a path four feet wide was soon made to the head of the first fall. But we were by no means through the canyon yet; for a mile more the river was confined between perpendicular walls of rock up which there was no climbing, and we had to seize on every "coign of vantage,"[38] narrow ledges of rock, banks of ice and snow clinging to the edge, bridges from one huge boulder to another; with the dark water boiling and foaming at our feet, ready to engulf anyone who made a false step. But the good ice was reached at last, and the party pushed on, well pleased at having surmounted so formidable an obstacle. Our joy, however, was of short duration, for once fairly launched in the mountain range [see Figure 3], canyon succeeded canyon, and the bed of the river became so full of boulders that progress was reduced to a minimum. About this time too (the beginning of February), the weather was very stormy and the falls of snow were frequent; the snow-shoeing become very laborious, and everyone's spirits depressed in consequence. Several moose showed themselves on the river, but neither the time nor the inclination for hunting (and perhaps the lack of necessity) induced us to go after them, so they were allowed to trot off in peace. Great numbers of ptarmigan passed over our heads in some of the canyons, but as the shot-gun had long ago been voted a nuisance and left at Fort George, they approached with impunity.

So great was the depth of snow here, that several times when standing on the blankets in camp (the snow having been shovelled out down to the moss) we could not see over the edge of the hole in which we were; and the wood pile was frequently overhead. But on the river itself the depth did not exceed two feet or two feet and a half; into this, however, the snow-shoe would sink a good foot, and coming up with a small avalanche on the toe at each step, caused many blisters and occasional *mal de raquette.*

The valley soon took a sharp turn to the north-east and entered the main range, while the river decreased in size, dwindling down to a mere creek tumbling down the mountain side. Here we left the dogs, and Hanington and I, with a couple of men each, did what climbing we could to discover the source of the river, each taking a different branch. But they both terminated in the same way, a small muskeg or swamp, of a semicircular form, surrounded on three sides by high bare rocky peaks, between which the long, clear blue line of the glaciers was painfully apparent. A return to camp was all that could be done – evidently no pass this way – and a long discussion over many pipes did not much help matters. As Hanington said: "We seem to have got to the back of the north wind," and I reluctantly gave up the idea of any more exploration in that direction. We were certainly in the heart of the mountains and would no doubt have admired the magnificent scenery under any other circumstances (out of the window of a Pullman car,[39] for instance) but the feeling of disappointment was too strong just now; scarcely even allowing us to take notice of the gambols of the "Bighorns" a thousand feet above us, who could be discerned through the field-glass taking stock of the intruders, and strutting up and down with a challenge, as it were, to scale the glacier and meet them on their own ground. There yet remained a possibility of getting through by a more southerly pass, and this might be reached by going up the south fork of the river, the mouth of which we passed a few miles above the cache. The sleds were accordingly loaded up, and we returned down river to the

"Forks" where we camped on the 12th February. Here we rested a day; and as the number of the party might now be reduced, there being a smaller amount of supplies to carry, two of the Indians and two of the hired trains were sent back to Fort George. Both dogs and men felt the good of a day's rest, after a month of incessant hard work.

The returning party carried with them our best wishes, and a letter to the Chief in Ottawa,[40] explaining the position and announcing my determination to make another attempt to find the Smoky River Pass, by following the south fork of the north branch of the Fraser River.[41] A small hand-sled having been made (by cutting down one of the large ones) three men started off with their blankets and a week's provision on it, the other three to follow a day later with their dog-trains. By this means it was hoped that a good track might be secured, and the work be made easier for the dogs; the trains, owing to the reduced number, being now loaded as heavily as at the start; and although the plan worked well for a week, the mild weather and almost total absence of frost at night, caused it to be abandoned at the end of that time.

Scarcely had we lost sight of our camp at the "Forks" when we came to a more mighty canyon[42] than any yet encountered, which necessitated a detour of about three miles overland to avoid it. In attempting at first to get through the canyon and thus avoid a portage through the woods we went over some very doubtful places, at one of which the rocks were overhanging to such a degree that Hanington had to take off his snowshoes (he going first) and creep along a ledge on hands and knees for fifty yards, while just beyond this, a fall (not very lucidly described by an Indian being "high all-the-same one stick") put a stop to any chance of getting through the canyon. Returning along the ledge, part of the snow slid away, but Hanington successfully imitated a limpet, clinging to the rock until a pole was held out to support him past this somewhat dangerous spot. In grasping the pole, however, he let go one of his snowshoes, which whirled away down stream and was given up for lost, when a sudden turn of the eddy brought it to the surface near enough to be fished out with the pole. At the south end of the portage the descent to the river was very steep, and with only one driver to each train, their downward course could not be confined to ordinary speed. The usual method of "putting on the brakes" by turning the sled on its side, and sitting on the curved bow was of no avail here, for in attempting it I was hurled to one side, and the whole train went pell-mell to the river, fortunately without doing any more harm than the breaking of a few traces. But Hanington devised a cunning plan, and "anchoring" his sled by the tail-rope to a tree was enabled to lower it gently for a short distance. When, however, he let go to change the rope from one tree to another, it became unmanageable and the whole concern started on its downward career, promising a repetition of my descent; but scarcely had the sled got abreast of the dogs when it sheered off to one side of a small sapling, they running or rather rolling on the other. The sapling bent, and the impetus carried the whole train out on it about 20 ft., the dogs hanging by their traces and just counterbalancing the sled, and swaying up and down in most ludicrous plight. A few blows with the axe set them free, and the river was reached without further mishap. The water had overflowed the ice in many places above the canyon, and this impeded our progress very much, as the bottom of the sleds had to be scraped every half mile to get rid of the slush sticking to them, which would soon have turned to ice. The sight of one small bit of clear, glare ice was hailed with a shout; even the dogs seemed to enter into our feelings, and set off at a scamper to cross it. But it was like a mirage in the desert, only meant to deceive, for no sooner did the weight of the sled come on it, than in it went, dogs and all – the ice proving to be no more than a quarter of an inch thick, and probably only frozen the night before. The water, however, was only a couple of feet deep, so they were easily fished out again. On the banks we saw several marks of old chopping, and at one camp found a very old axe, like those made years ago by the blacksmith, at York Factory,[43] on Hudson's Bay. This was cheering to the whole party, as it seemed to prove that we were on the right road to the desired pass. Old Indian stories tell of the time when the Crees used to cross the mountains here, and even bring horses as far down as the first big canyon.

The valley of this branch is very similar to the one first followed; and at about the same distance from the Forks it also turned off to the north-east, and we entered the Rockies again. Here there was a great deal of open water, caused probably by the extreme mildness of the weather for the last few days. There occurred here one of the most sudden changes we ever experienced; going to bed one night with all our available clothing on, and the thermometer at forty-two degrees below zero, we were awoke next morning by the pattering rain on our faces, and found the temperature had risen to forty – a change of eighty-two degrees within eight hours. We were enervated by this, as it appeared to us, sultry heat, and the dogs went along panting, with outstretched tongues. Our snow-shoes also gave way, being thoroughly water-logged, and half a day had to be devoted to repairing damages. During this afternoon, Alec noticed one of his dogs, a fine bull-pup named "Captain," wandering about in an unsettled way, but he finally brought himself to anchor on top of the wood-pile (turning round twice as all dogs do before lying down, because, I suppose, "one good turn deserves another") a post that he ever after successfully held against all comers; and he even went the length of plainly intimating that he desired to be fed in no other place than that. As "Captain" was a great favourite, his very reasonable request was acceded to, though not without sundry pitched battles between himself and the Husky (or Esquimaux) dogs, who seemed to object to any partiality.

The entrance to the Pass is very grand, being guarded on either side by high pyramidical peaks[44] towering two to three thousand feet above the valley and covered with perpetual snow [see Figure 4]. To the most prominent of these points we gave the name of "Mount Ida," and it was here we saw one of the most magnificent of the many fine glaciers along the route;[45] it could not have been less than a mile long, and five hundred feet thick at the face; while it was of such a transparent blue that we could almost imagine seeing the rocks underneath and through it. Just when I had chosen a place to camp, a roll as of distant thunder was heard, and a mighty avalanche seen rolling down the mountain side just above us, the masses of ice and rock chasing one another and leaping from point to point as if playing some weird, gigantic game. While we were discussing the probability of its reaching us – which however was strenuously negative – down came one huge boulder as though making directly for us; but being turned aside by the trees as it crashed through them, plunged into the river a chain in front of the dogs, who appeared puzzled to account for its sudden arrival. It was of limestone and about ten feet in diameter. We did not camp near that spot.

The next day the stream began to rise rapidly and become much smaller, a good deal of open water drove us from the ice into the woods; and finally a sudden termination of the valley and the usual small stream trickling down the mountain side, showed conclusively that no practicable pass existed here. But the weather was fine, cold and exhilarating – and I decided to push on to the summit, if there were such a thing possible. Abandoning the creek, we climbed a couple of thousand feet to a lake[46] whose dimensions were shrouded in mist – all we could tell about it was that it is the head of one branch of the river we have followed up. Leaving the camp near the lake, Hanington and I went ahead four or five miles, and passing through as many lakes nearly all at the same elevation,[47] hailed with joy the appearance of water running to the east, and returned to camp to tell the welcome news; at the same time making known my intention of pushing on towards Edmonton, rather than to turn back after having gone so far towards it. Everyone being anxious to see "the other side" we were off at gray dawn and well across the large lake before the rising sun gave a splendid view of the surrounding country. The Lakes lie in a long deep gorge running due east and west through the mountains, about a mile wide, and perfectly straight for seven or eight miles [see Figure 5].[48] Having discovered the exact spot at which the waters divide, several trees were blazed, and having marked one of them in a conspicuous position, as the "Boundary between British Columbia and the Nor'-West Territory,"[49] with our names and the date, we started "Eastward ho!" with more satisfaction than we had felt for many a day. At the little stream issuing from the east end of the lakes we took our first drink of water flowing to the Arctic Ocean, and supposing we were at the head of Smoky River, we christened the peak which guards this end of the pass, "Smoky Peak."[50] The stream soon became large enough to travel on, and with such an evident down grade as to call forth allusions to *facilis descensus Averni*;[51] and by inverse ratio, the lower we got the more our spirits rose. The most curiously noticeable fact was the rapid increase in the size of the river, which, at the end of the first day's travel on it, was already a couple of chains wide, and this without the visible addition of any branches which would help to swell the volume of its waters.[52] Early the second morning, we met with a check; Hanington and I were ahead when, on turning a sharp bend in the river, an immense abyss yawned before us, and we stood on the very edge of a fall which proved to be two hundred and ten feet high, and over which, had the morning been at all misty, we would probably have walked [see Figure 6].[53] There was no sound of falling water to give warning of the dangerous proximity; and it afterwards appeared, when looking up from below, that the whole party had been standing on what was merely a projecting cake of ice and snow, not more than a couple of feet thick. The left bank appeared most favorable to make a portage on, and we had to go back a short distance to climb the hill on that side. I took the first opportunity of descending again to the valley of the river, sitting as usual on the heels of my snowshoes, but taking some rather ugly and unforeseen jumps over sundry little bluffs near the bottom, and finally landing minus mitts and cap, full length in the open water, fortunately only about a foot deep. The others continued the portage further, as it was impossible to descend near that point with the dogs; but they had eventually to come down a place very nearly as steep, where one of the sleds broke away from the driver, and coming in violent contact with a log in its downward career, made a sandwich of the unfortunate dog nearest the sled, and broke the "nose" (or turned-up bow) into a dozen pieces, besides damaging the harness. This was our first serious calamity, but, the dog excepted, everything was set straight in a couple of hours – the poor animal was past all care when the sled struck. A trivial incident like the death of a dog (and especially such mongrel curs as some of ours were) would not affect one seriously in a civilized community; but it cast quite a gloom over our little party, and even the dogs looked at one another, as who should say, "It may be my turn next!"

At the foot of this canyon we found ourselves fairly out of the mountain range; the few spurs that follow down each side of the valley are low and timbered to the summit (or rather were, for the whole country has been very recently burned over), and the bare rocky peaks were soon lost sight of behind us. The next day we saw marks of Indian chopping, and camps, apparently of last summer; and here, for a distance of twenty miles, we noticed the almost total absence of snow, a phenomenon said to occur all along the eastern base of the mountains. At one of our camps there was not more than two inches of snow anywhere in the vicinity. Our rate of travelling improved in consequence, and we made one big drive; but next day the old state of things returned, and the snow soon reached its average depth of two and a half feet, making the walking terribly heavy. This told on the dogs, who were getting tired out with their incessant hard work, and we had frequently to leave the sleds standing and the whole six go ahead to break track, then three return to bring on the trains, and still find the snow so soft, even after the passage of nine pairs of snowshoes, that the poor animals would wallow through it up to their bodies. It soon became evident this sort of thing could not last much longer; it was beginning to tell on the men as well as the dogs; and another cause for uneasiness suggested itself, "what if this be not Smoky River at all, but some other branch of Peace River, which will take us away, goodness knows where, if we follow it?" For we knew by our latitude, obtained by observation, and our approximate longitude, calculated by dead reckoning from the track survey we were making, that our course to strike the Athabasca River and the country we wished to explore between it and the Saskatchewan River would be about south-east, while we were now travelling at right angles to this course, or north-east. I could not, however, abandon the hope of the river shortly turning to the east, or even more in the desired direction, so we held on a few days longer. But scarcely a day passed when the dismal howl of the dogs did not announce to our unwilling ears that another of their number had dropped exhausted in his tracks; and it soon became very evident that we must put our best foot foremost in order to get through with safety to ourselves. At camp this night we saw a number of old Indian lodges, and marks of a horse-trail having been cut through the woods; this encouraged us to think we were on Smoky River, as Alec knew the Jasper Valley Indians had a trail by which they go in summer to it, but in winter we cannot find sure proof that it is a trail of pack animals. A long and earnest consultation, in which three different propositions were made: 1st, to assume we are on Smoky River, and to follow it to Peace River and Fort Dunvegan;[54] 2nd, to go east to Fort Assiniboine,[55] on the Athabasca; and 3rd, to go south-east to Jasper House, ended in the adoption of the latter; and the following day, finding the river turned still more to the north, orders were given to camp early, and a suitable place chosen to build a cache in which to leave everything that could possibly be spared. Going ahead a couple of miles to look for a good place to leave the river, we came across a very old and indistinct snowshoe track coming down on the river and after half a mile leaving it again, but without the least vestige of a track in the woods. This part of the country is evidently not much visited in winter; the scarcity of game would account for this, for we have seen absolutely nothing since leaving the mountains. We certainly expected to get deer or moose to eke out our stock of provisions, now becoming very small, but not a single one has been visible lately. Next day, the 6th March, we remained in camp to take a much needed rest, and to make various repairs. I determined, if possible, to take one train to Jasper House, and Alec's was selected for that purpose; the other two sleds and their harness, together with superfluous clothing and instruments, were placed in a small log hut, six feet by four, and three feet high, built for the purpose, and the names and date marked on surrounding trees.[56] Knowing the extreme difficulty of getting a loaded train through the woods, each one was to carry his blankets and share of the provisions, while the salmon were made into little packs and divided among the dogs, who would surely be able to get along with these small loads (not more than fifteen pounds apiece). The following day we started early, and by this must be understood a couple of hours before sunrise, our usual time for departure from camp being as soon as we could see to put one foot before the other, which necessitated rising at four o'clock every morning throughout the whole winter.

Going a couple of miles down the Smoky River we turned off to the south up a small creek,[57] being anxious to keep as long as possible out of the woods; but so much time was lost in following its various turns and windings that we struck off to the south-east and pushed on, over hill and dale, regardless of anything but progress on the course laid down. We soon got into a terrible thicket of small black pine, growing so close together that we could scarcely force a passage through them; and at sun-set we laid down tired out and disgusted at having only made seven miles. Another couple of miles in the morning brought us to a river, the counterpart of the one we left, and which is probably its south branch.[58] The high bluffs on the south side looked so forbidding that we went a couple of miles up the river, till we found a small creek coming in from the south-east, up which we turned.[59] Looking for dry kindling wood in a drift pile of brush and trees at the mouth of this creek, Alec called our attention to what seemed a veritable saw-log, evidently cut by a white man (the Indians do not tackle anything over six inches diameter) and which must have drifted down from above. This puzzled us considerably. Was it possible we were on the Athabasca? Common sense said no; but then, how to account for the saw-log? If it be the Athabasca, then by keeping on our course we must soon strike the Macleod River;[60] and the river we followed down from the mountains must be either Rivière à Baptiste or Old Man's River,[61] and not the Smoky River as we thought. But we pushed on to the south-east, and only discussed these abstruse questions over the campfire. The country was very broken, and consisted principally of long high ridges crossing our course at right angles, and covered with small pine of second growth. The frequent "ups and downs" were hard on the poor dogs, who were very weak, and fell exhausted daily; in order to spare them any more suffering, the stragglers received a *coup de grace* from one of our revolvers, and the others, "closing up" continued the march, only howling a requiem over their dead companions round the camp at night. From the top of one of the highest ridges, a perfect "Hog's back," we caught sight of a deep valley at our feet, and the Rockies fifty miles away to the south. This must be the Athabasca, and we hastened down, eager to reach a known point; but only to be disappointed, for it turned out to be a vast muskeg, nearly treeless, and from which we got a good look at the mountains away to the northwest, almost as far, we imagine, as "Smoky Peak." We must surely come to some water running the other way soon, which will be a sort of guide to us. Near camp to-night I found signs of a trail, a few trees having been blazed, but it did not appear to run in the right direction for us. Another high ridge loomed up in front, and surmounting it after much hard climbing, traces of the trail were again found, with old Indian camps and the head waters of a river running to the south-east. Surely we were now approaching the Athabasca! We plucked up heart and made a good day's march. But the blazes, at first easy to follow, become indistinct and finally lost before night, and when orders were given to camp Hanington and I started out in different directions to look for them, leaving Alec and the Indians to make the camp. On my return I found the Indians in a mournful state of despair, declaring they were lost and would never see their homes again, and weeping bitterly. It took a great deal of persuasion to set them on their legs again, and had there been any possibility of their running away there is but little doubt that their fears were so worked upon that they would soon have availed themselves of it. But they knew their only chance of coming through in safety lay in remaining with the party, and they submitted to our arguments, though we found it somewhat difficult to use persuasive eloquence where we were not quite sure of the soundness of our own reasoning. The river, as usual, began to turn off to the north-east, so we decided to leave it and follow the old south-east course, which has so far led us into no great difficulties. The dogs decreased rapidly in number and size; a great favourite of mine, one of the Cariboo dogs, called "Buster" – probably a contraction of Filibuster – could not be coaxed away from the camp fire this morning, but no one had the heart to put an end to him, so he was left to his fate, not without many regrets. By this time we expected to have been near the mountains seen some days ago, and possibly may have been but a thick mist shrouded everything for a couple of days and we groped along almost in darkness. But one bright morning the rising sun dispelled the mist, and from an elevated and burnt side hill on which we were travelling, Alec caught sight of a, to him, well-known feature in the landscape, the "Roche à Miette"[62] whose peculiar and distinct profile was plainly visible about twenty-five miles south of us [see Figure 7]. This mountain is opposite Jasper House, at the eastern end of the Yellow Head Pass,[63] and the sight of it was an immense relief to the minds of the leaders of the party, since it was from the Hudson's Bay Company's post there that we expected shelter and supplies, the latter having now reached very small proportions. The packs were thrown off in the snow, and we took a long rest and smoke – the feeling of security after the anxieties of the past month was too pleasant to be rudely disturbed, and even the stolid countenances of the Indians lighted up at the thought of a good feed and a respite from their incessant labours. But sitting on a log would not advance us much, so we marched off again, and getting on the ice of three or four small lakes[64] made good time towards our goal [see Figure 8].

A bluff precipice intervening soon shut out our view, and to avoid it, we turned away to the left, crossing a high and heavily timbered hill, on the eastern slope of which we camped, with pleasant anticipations of returning to the land of the living on the morrow. But after supper Alec was seen stealing quietly away from camp, and being closely questioned on his return, admitted that he had gone to take another look at the "Roche" by moonlight, to assure himself that he was not mistaken. The bare possibility of such a thing alarmed us, and the evening did not pass as cheerfully as it begun. One thing was very evident, if that was not the Roche à Miette and the Athabasca in this hole at our feet, we might as well give up the hope of ever finding either, and the prospect was not inviting. But we slept well, nevertheless, for the clear bracing air, plain (not to say meagre) diet and constant hard exercise, ensure that.

About three miles from camp next morning we found ourselves on the benches overlooking the long-sought river, and it became a perfect scamper who should reach it first – *mal de raquette* was forgotten, (though it is generally a pretty attentive companion) and the half-starved dogs staggering along after us, joined in the enthusiasm with the most feeble of barks. But the effort was too much for them, and one more faithful servant dropped in his traces a few yards from the river bank. Ascending the river a couple of miles we came to the "Lac à Brûlé"[65] where the ice was almost glare, the snow being blown off by the furious winds that rush down through the Pass like a funnel; and we travelled without snowshoes the first time for three and a half months [see Figure 9].

The eight miles up this lake was soon got over, and arriving at the Fiddle River Depot (built by Mr. Moberly)[66] we were cordially received by the Iroquois Indians camped there [see Figure 10]. An immense dish of boiled rabbits set before us disappeared in quick order, and after this good meal we were more reconciled to hear the Company's post at Jasper House was abandoned. What was now to be done? We were at least ten days' journey from Lake St. Anne's, the nearest post we could depend upon, with only about two days' supplies remaining. The Indians could not give us anything, so we seemed to be in a tight place. But a long talk with an old squaw, who spoke very good French, ended in her promising to get everything that could be spared for us in the way of provisions, and the opportune display of a little money, raised the *auri sacra fames*[67] to such a pitch that, early next morning by collecting from the various lodges round, we scraped together some sixty pounds of dried deers' meat; and as there was no immediate prospect of starvation, a halt was ordered for the day. I and one of the Indians rode up to Jasper House, about seven miles by the trail, where a quarter of mutton (mountain sheep) had been cached. There was nothing at the store but a little powder and shot; so we returned to the Depot, and the afternoon was spent in dividing the provisions into packs. The number of dogs was now reduced to seven, and as they were too weak to travel – besides not being able to spare them any food from our scanty supply – it was arranged that the Indians should take care of them until they could be turned over to the Company or some of our own people. Having bought some moccasins and rewarded our kind entertainers, we shouldered our packs and turned our faces towards Edmonton.

The wind blowing as usual directly down Lac à Brûlé was this time in our back, but the ice was so glare that we could not keep our feet; and after staggering and creeping along for half a mile, we had to put on our snowshoes and skirt the shore. On the Athabasca the going was good, and we were not long in making the twenty-five miles, to the point where we intended to leave the river. Keeping to the north of the line run two years ago by Mr. Moberly,[68] we marched nearly due east to the Macleod River; but our progress was very slow owing to the great depth and softness of the snow as well as the dense thickets and tangled brûlés[69] we had to force our way through. We were fortunately not encumbered with dogs, or we would have spent still more time in choosing a passable road for them. From what is called the "Macleod Portage"[70] the last view of the Rocky

Mountains was obtained, and few among us were loath to turn our backs on the scene of toil. The view from the east end of the portage is very fine; a panorama of immense extent lay at our feet, and the horizon for a distance of fifty miles was bounded by the lofty crests and snowy peaks of the "backbone of the Continent" rendered more beautiful than ever by the rosy hues of the rising sun, and becoming more and more interesting to us as we left them in the distance, and shook the snow from off our feet against them. Shortly after going down on to the Macleod again, we met the Hudson's Bay Company's outfit going to Jasper House to trade with the Indians we saw there; and from them we were fortunate enough to get a little tea (ours being exhausted), and a few pounds of pemmican to recruit our very scanty larder. But our pleasure at having their track to travel on and thus save our weary legs was dashed by learning them to be eleven days out from Lake St. Anne. A careful division of provisions that evening gave us four days more, or perhaps five, if we could manage with less than a pound a piece each day; and we did not like the thoughts of what we were to do during the other five or six days. But there was nothing for it but to push on and hope for the best, so we followed the track two days down the Macleod. Here it became completely snowed up and overflowed; besides which, I thought it a round about way to follow the river, with all its windings, so far; and accordingly struck off due east towards Dirt Lake,[71] which we were fortunate enough to fall in with next evening.

A curious sensation of numbness now began to take hold of our limbs, with an unwillingness, or rather inability to push one snowshoe before the other after lifting it up; this gave us the appearance occasionally of "marking time" and would no doubt have been amusing to a well-fed bystander; but to us it was no laughing matter. Frequent cramps in the hands, caused probably by the pressure of the pack-straps on the shoulders, also added to our discomforts. A couple of rabbits opportunely appearing near camp gave us an apology for a breakfast; and the evening of the third day after, we reached the Hudson's Bay Company's post at Lake St. Anne. The intervening time was probably spent in a sort of mechanical progress, for nobody seemed to have any very distinct ideas, except on the subject of looseness in the region of the waistband. We were very kindly received by

Mr. McGillivray,[72] the officer in charge, who set us down at once to a good meal of white-fish and potatoes; and, after the manner of starving men in general, we ate a great deal more than was good for us. There never was a more welcome riddance of a burden than when we threw down our packs and took off our snow-shoes at Mr. McGillivray's door, for although the loads did not probably exceed thirty pounds each, they felt, on our weak shoulders, like a hundred. The next day was given up to much needed repose; and there being a beaten road from here to Edmonton, I arranged with Mr. McG. to furnish a couple of horses and sleds to convey us there.

We made the fifty miles to Fort Edmonton in a day and a half, and were hospitably received there and entertained by Mr. Hardisty,[73] the gentleman in charge of the district. But our four days' rest was not very enjoyable; we all suffered much from cramps in the limbs; and the sudden change from semi-starvation to a liberal diet brought on an attack of dysentery, and it was some days before we completely recovered our strength. As the two Indians belonging to Stewart's Lake[74] have to return there in the spring, I made an arrangement with Mr. Hardisty to keep them at his Fort in the meantime, and to give them a pack-horse and provisions as soon as the snow goes, to return by way of Jasper House and Tête Jaune Cache.[75] Some horses of Moberly's party having been left at the Fort, we obtained ten of them, and procuring flat-sleds, buffalo robes and some necessary additions to our wardrobe; and taking charge of a "Packet" of mail for Fort Garry,[76] we started east, with Jack Norris[77] as guide, on April 7th. Although our horses were very poor – as was the case with the stock both of the Company and the surrounding farmers, owing to the insufficiency of last year's hay crop, and the lateness of this spring – many horses and cattle having starved to death – we made the eighty miles to Victoria mission in four days [see Figure 11]. We were lucky enough to get a little barley and hay there; and, a very sudden thaw coming on, we remained with Mr. Adams,[78] the gentleman in charge of the Company's post, for two days. The horses benefited by the rest and feed; and as we imagined the winter to be at its last gasp, and that the snow would soon leave us altogether, we concluded to exchange the flat-sleds for carts, to be obtained at Fort Pitt; using in the meantime the pack-saddles brought for this purpose from Edmonton.

We accordingly started with five packs and five light (or unloaded) horses – and to avoid the snow in the woods near the Fort, followed down the Saskatchewan on the ice. But as this was knee-deep in water, we could not bear it after the first day; and climbed the bank again where we, with our snow-shoes, made much better progress over the drifts than the horses did through them. But the snow gradually got less and patches of bare ground appeared; being seized upon with avidity by the half-starved animals, glad to get a few blades of grass without "pawing" (digging away the snow with their hoofs) for them. We hung up our now useless and tattered snow-shoes on a tree, with suitable inscriptions. Water began to trouble us more than snow, for all the small creeks were breaking up, and invariably overflowed their banks. In some of them, the old ice (though under water) remained sound enough to carry us safely over; but in more than one instance – and notably at "Dogrump Creek"[79] – the whole valley, a quarter of a mile wide, was covered with a rushing torrent. Riding out to explore this waste of waters (up to the horse's shoulder on the flats) I suddenly plunged into the channel, but in spite of the shock of the icy water up to the neck, – the horse swimming low – was able to guide him across and effect a landing on the other side. The rest soon followed this involuntary example, and hauling the pack-horses across with a tow-line reached the east bank in a cool condition. A keen north wind soon coated us with ice, and as we had to go a couple of miles before finding any firewood, we were tolerably numb and shaky. The thaw, which we so confidently anticipated, came on but slowly; so that there was still a foot of snow when we reached Fort Pitt, six and a half days making the one hundred and twenty-two miles from Victoria. Mr. McKay[80] kindly placed a room at our disposal, and fed us with his best. We here left our pack-saddles, taking in their place two carts. The ice on the Saskatchewan had broken up on the 18th, and the river, rising higher than usual, reached almost to the gates of the Fort. Wild fowl arrived daily; and, although the depth of snow seemed to render carting rather premature, the signs of spring were so many that it could not be long before it came. At the first hill we came to after leaving the Fort, our troubles began; for some of the horses had no idea of being in harness, and no amount of persuasion would bring them to collar – harshness and kindness were tried alike in vain, and for some days we had to change horses at every steep place. The most incorrigible ones were subdued by the half-breed remedy of tying a rope from the shafts of the cart to the horse's tail, and by means of this novel tandem we generally succeeded in getting out of the worst places. But the drifts caused more delay than the hills; for the carts would generally run a little way out on to them and then settle quietly down to the axle; in many places we had to march up and down more than a hundred yards, jumping on the crust to break it and make three tracks, one for the horse and one for each wheel. The light animals were also driven through first to help break trail. After a few days, however, the "baulky" horses became knowing and refused altogether to venture into a drift unless unharnessed; so that we had several times to pull the carts through by hand ourselves. As most of the creeks were now quite open and very high, we rafted across them; not being in favour of any more swimming matches. The rafts used by all travellers on the plains are after one pattern; an oil-cloth or tent stitched over a framework made, according to circumstances, of willows, waggon-box [*sic*] or cart-wheels. The latter mode we adopted; laying two wheels side by side, overlapping till the rim of one touched the hub of the other, and lashing them firmly in this position. A tarpaulin was then spread on the ground, and the frame laid upon it, the ends and sides being turned up and tied to the rims. In this manner we were able to transport all our impedimenta across any stream where the tow-line would reach, in three or four trips. But winter seemed to have repented its departure, and returned in the shape of heavy frosts every night, which made our work of cutting through the drifts exceedingly labourious, and progress consequently slow. The crust became at last so solid that for forty miles above Carlton we drove our loaded carts on its surface; and instead of avoiding the drifts as heretofore, steered for the biggest of them, knowing they would best bear us up. In this fashion we reached Fort Carlton [see Figure 12] on the 29th April, having taken eight and a half days to come from Fort Pitt, a distance of one hundred and sixty-seven miles. The Fort stands on the south bank of the river, and when we came down the road on the north side, we could see little else but immense blocks of ice piled many feet high on either shore. After much shouting and gesticulation – the Fort being half a mile from us – we had the satisfaction of seeing a Mackinaw boat push out, and, skilfully avoiding the floating bergs, come to our side. Hanington and I took the packet, and, crossing over, were well received and entertained by Mr. Clarke,[81] the officer in charge of the district. As it was now late in the day, the rest of the party camped on the other side.

The following morning, the horses were induced, after much persuasion, to trust themselves to the boat, and everything was safely crossed in two trips. We pitched our camp just outside the Fort, but Mr. Clarke insisted on our remaining in his house. The river broke up here a few days later than at Fort Pitt, and with more disastrous consequences. The water rose very suddenly and carried off the scow which was used as a ferry; and surprised some Indians, in the night, who were making maple sugar on an island twenty-five miles below; these poor wretches took to the trees, but as no help could reach them, dropped off exhausted one by one, till by daylight there were none left out of a band of a dozen. At Fort à la Corne,[82] a hundred miles lower down the river, the water stood four feet deep in the Company's storehouse, and all the goods had to be moved upstairs, the people themselves taking refuge in the hills at the back. No news had been heard from Cumberland House,[83] still farther down the river, and it was feared they might have suffered much, being situated in a low and flat region; but it afterwards appeared that this was the very cause of their safety, for the water, spreading over the surrounding country, lost the destructive effect it had when confined to a comparatively narrow channel, and passed harmlessly by.

For three days after our arrival, a keen north wind delayed the departure of the snow; but the beginning of May was warm and genial, and we prepared to take the road again. Some fresh horses were procured with great difficulty, and Alec brought some barley from Prince Albert's mission,[84] fifty miles down the river; the country being reported burnt from here to Ellice,[85] and feed consequently scarce. On the morning of the 5th May we climbed the hill behind the Fort and set our faces towards the rising sun. The snow was nearly all gone, and we easily avoided the few remaining drifts. We reached the South Branch of the Saskatchewan[86] the same afternoon, and spent four hours making two trips across with the scow, as a strong south-west wind (down river) necessitated the hauling of the scow a long way up on our side to ensure making a good landing on the other. At the French half-breed settlement here the people were driven out of their houses by the rising waters, which seem to have been bigger this spring than for many years past. The grass having all been burnt off last autumn, gives the country a cheerless aspect; and we had to go to the margins of lakes or swamps to find any feed at all for the horses; but they, pushing on at the rate of thirty miles a day, with the characteristic endurance of Indian ponies, did not seem to feel the hardships of the trip as much as we had expected.

About forty-five miles from the South Branch we passed the "Spathanaw," or Round Hill [see Figure 13],[87] a conspicuous feature in the landscape; with a wooden crucifix on its summit, said to have been placed there by a worthy Bishop who spent Sunday at its foot.

Not far from here, a road branches off to the south west, crossing the South Branch[88] above where we did, and here we met with the first appearance of civilized usages – a finger post indicating Gabriel Dumont's crossing [see Figure 26].[89]

The latter statement was especially interesting; but we took it for granted that these for whom it is intended can make more out of it than we could; so we went our way and reached Touchwood Hill Post[90] on the evening of the 9th. Here we left one of the hired horses, and as the others were already showing decided symptoms of "giving out" we had to continue the journey on foot, without even the occasional rest of a mile or two in the saddle, the animals having to be spared for use in the carts. But, wearing only moccasins, we found the unaccustomed exercise beginning to tell upon us at the end of a hundred miles, and by the time we had accomplished fifty more, were so footsore that we were quite ready to avail ourselves of a seat in a cart for half an hour when the half starved horse seemed in a livelier mood than usual. A couple of days above Fort Ellice, we met two travellers, by name Livingstone and Fraser, footing their way towards the mountains, thence intending to strike for the Cariboo mines. They jogged along in primitive style, unencumbered by either blanket or provisions, carrying only a spare shirt, a gun and some ammunition. To save the necessity for a blanket, and also to avoid the heat, they slept by day and marched by night. On the evening of the 14th we camped at the mouth of the Qu'Appelle River[91] and crossed over to Fort Ellice[92] early next morning; ten days making the three hundred and sixteen miles from Carlton. Here we received every kind of assistance from Mr. McDonald,[93] the gentleman in charge; who, having no available horses of his own, endeavoured to replenish our scanty stock by hiring or purchasing for us from others. We made but a short halt, crossing the Assiniboine River[94] on the scow after dinner – the bridge having been carried away by the freshet – and pushed on a dozen miles to the east over a very good road. In saying goodbye to Mr. McDonald, we parted with regret with the last of a number of gentlemen, officers of the Hon. Hudson's Bay Company, who have shown us every kindness and extended a ready hospitality on every occasion we have come in contact with them [see Figure 14]. To most, if not all of them, we were personally unknown; but it was sufficient to say we were in need of help, to ensure at once their best endeavours on our behalf.

We were delayed until ten o'clock next morning, as our horses had seen fit to rejoin their companions near the Fort, but we got past Shoal Lake[95] before camping time. In crossing the Little Saskatchewan River[96] we had a good deal of trouble, the water was very swift and high, being above the horses' backs. The load had to be piled on an improvised rack on top of the cart body, and by an ingenious combination of tow lines, the horse swimming and the cart afloat, they were safely piloted across. This was our last excitement, except the breaking of an axle against a stump a few miles farther on, and we soon reached the flourishing settlement at the third, second and first crossings of the White Mud River[97] where the farmers were busy with their spring occupations, but not over-sanguine of success, owing to the annual scourge of grasshoppers, which has hitherto turned this fruitful colony into a barren waste.

Passing Portage la Prairie[98] on the 19th, we reached Winnipeg on the 21st May, having been five and a half months on our trip. At White Horse

Plains[99] we met a gay cavalcade going westward; it consisted of Mr. McLeod[100] and his two survey parties, just starting for Edmonton and the Rocky Mountains, and their shining boots, glittering spurs and well-groomed horses contrasted with our battered and weather-worn appearance. But we could afford to suffer by the comparison; they would soon be as ragged as we were, and all their troubles were before them, while we were just reaching the goal, pushed forward to over many a weary mile of mountain and plain, and could take our well-earned repose in the happy consciousness of having fulfilled the task allotted to us, and earned the approbation of him we are proud to acknowledge our Chief.[101]

EXPLORATION OF SMOKY RIVER PASS, 1874–5.
TABLE OF DISTANCES

From Quesnelle Mouth B.C.	Actual Days'	Snow-shoes with dogs	Snow-shoes with packs	Horses flat sleds	Horses packs	Horses carts
To Fort George	10	135				
Forks on North Fraser	7	98				
Salmon Cache & return	6	90				
Head of North Branch & return	13 ½	100				
Entrance to Pass	6	48 ½				
Summit	4	34 ½				
Smoky River Cache	8	87				
Fiddle River Depot	11		108			
Jasper House & return	1		14			
Lake St. Anne	13		217			
Edmonton	2			50		
Victoria	4			80		
Pitt	6 ½				122	
Carlton	8 ½					167
Ellice	10					316
Winnipeg	5 ½					220
Total	116	593	339	130	122	703

Days 16 Miles 1,887
Average per day, miles 16.26

CHAPTER 2

C.F. Hanington's Journal

JOURNAL OF MR. C.F. HANINGTON FROM QUESNELLE THROUGH THE ROCKY MOUNTAINS, DURING THE WINTER OF 1874–5[1]

Archivist Douglas Brymner prefaced the presentation of the journal with these words:

The difficulties encountered in the exploration for the line of the Canadian Pacific Railway through the Rocky Mountains may be guessed at by reading the official reports, but cannot be fully understood, the bare results being given and very little notice taken of the sufferings of the men engaged. At Note C is given the journal of Mr. C.F. Hanington, addressed in the form of letters to his brother, the Rev. E.A.W. Hanington, New Edinburgh, who has presented it to this branch, together with a map of the route, extending from Quesnelle, in British Columbia, to Lake of the Woods. As the journal only gives details of the journey to Edmonton, and that the rest of the route is comparatively well known, only the part of the map between Quesnelle and Fort Victoria, a little beyond Edmonton, has been lithographed [see Figure 15].

In Mr. Sandford Fleming's "Report on Surveys, &c, on the Canadian Pacific Railway," published in 1877, is a short narrative by Mr. E.W. Jarvis, at the head of the party of which Mr. Hanington formed part. The narrative by Mr. Jarvis is very interesting, but it, of necessity, fails to give those minute details and personal feelings which are to be found in Mr. Hanington's journal, the one being a narrative drawn up for publication, with all the reserve which that fact implies, and the other written from day to day unreservedly and whilst every impression was fresh and the most trifling incidents fully remembered. Both narratives should be read together.

QUESNELLE,[2] B.C.,
December 6, '74.

MY DEAR EDWARD, –

We have finished our season's work in the way of line running and have come down the Fraser River to this place, a town, as it is called.

After our line was finished I took two Indians and a canoe and made a cache[3] up the North Fork of the Fraser,[4] where we go next winter to explore a supposed gap in the Rocky Mountains, known as the Smoky River Pass.[5] The North Fork joins the Fraser River some 60 miles from Fort George[6] and I went up the N. Fork some 30 miles till a canyon prevented my further progress. I then cached the bacon and flour and returned down stream to this place. Quesnelle is (by the River) 82 miles below Fort George and there are two canyons[7] in that distance, both of which we passed through without difficulty.

The party (Divn. M.)[8] went down by stage and steamer to Victoria while Jarvis (in charge) and I are waiting here for cold weather. When the river takes a notion to freeze we start at once. Quesnelle is a queer sort of a place with a strange mixture of several kinds of people. Most of the inhabitants have been miners and go into other business when their coin runs short. The Hotel is kept by Brown and Gillis [see Figure 16],[9] who do things in first class style and charge $3.50 per diem for doing it. Drinks, beer or otherwise, 25 cents per glass, very small glasses. Gillis is a native of P.E. Island and a good fellow he is. As Jarvis is also a P.E. Islander and I a Blue Noser,[10] we are great friends of Gillis. The butcher in this town is also from the Lower Provinces, being a Haligonian;[11] his brother is organist in one of our churches there and poor Mike (Hagarty)[12] has gone into the meat business having failed in the mines.

There are several stores here, Read's [see Figure 17],[13] Girod's[14] and Kuong Lee's[15] being the most important. Read is a capital fellow and keeps a lot of good cigars for his own and friend's use. Girod is a Frenchman and hot after money. Kuong Lee the Chinese firm do a very large business in all sorts of goods, they have on hand a lot of Green Ginger and several kinds of fruit which I had never seen before, but which I like exceedingly. Like the other merchants, they are very good at "setting it out" for their customers.

We are here only 60 miles from Cariboo[16] the great mining region of B.C. and the E. end of the stage road and telegraph line.[17] They are doing well up at Cariboo just now and each week's mail brings down more gold dust than I'd like to carry; it goes to Victoria and is there sold to the banks, who either sell it, or send it to San Francisco to be coined. I may add that the Victoria company took 1,100 ounces out of their mine last week and it has been doing nearly as well as that for some time. Gold is worth $16 per oz.

We have been here since the 20th October, and are getting very sick of it. We have a telegraph wire from the main office to the office of the hotel, and Jarvis and I practice a few hours each day; I have become quite an operator, and shall keep at it till I am a better one. We take a walk each day to keep our muscles in order, for though we make light of the coming trip, it is going to be rather a tough one. In the evening we either spend the time in Read's store with cigars and talk, or sit around the huge stove in Brown & Gillis' with our pipes, and listen to the yarns of the miners, who are always ready to tell a good one. I like Quesnelle very much indeed. It is slow enough, but the fellows are jolly and independent, and the grub good. The population is, I forgot to say, about 100, including Chinamen and Indians.

December 7th, '74. We have concluded to abandon the idea of going to Fort George by the river, and take the trail for it, though the latter is a good deal longer and very much the harder road, but the river shows no sign of being frozen, and as the season is getting on we must go. We had engaged an Indian (Johnny) and a young Red River quarter-breed, who has been in British Columbia some two years; his name is Alec McDonald. We had also bought two teams of dogs (8), and got the sleds loaded for a start tomorrow. The dogs look first rate, being large, long-haired and fat. One, "Chun," is a tearer; we bought him from an Indian, who had him muzzled to ensure safety. Alec and I got him and fastened a long stick to his neck, and started to take him to the hotel. Of course the 100 Quesnellites turned out to see the fun, and they made it lively for us, with advice how to treat a dog who wouldn't come where he was wanted. In the midst of it poor "Chun" got loose, and the way he cleared that sidewalk was a caution to dog fanciers; some of the people nearly got into the river in their fright, while Chun went off to the bush, where he was captured next day. Our dogs had made it rather uncomfortable for the people here, who prefer to sleep at night instead of being kept awake by the doleful music of eight good howlers. But you ought to hear train dogs sing to appreciate their feelings. My train is "Marquis" leader, "Cabree" 2nd, "Sam" 3rd, and "Buster" 4th. The dogs are harnessed one before the other, and fastened to the sled by traces only; I mean there are no shafts. I'll write you from Fort George, where we hope to be in a week or so.

Yours,
C.F.H.

FORT GEORGE, B.C.,
19th Dec., 1874.

MY DEAR EDWARD, –

I wrote you on the eve of our departure from Quesnelle and I now continue from that point. We got away from Quesnelle on the 8th about 12 noon with teams pretty well loaded with grub and other supplies. Ben Gillis "set it out for us," and the whole town turned out to bid us "God speed." They had a very exalted idea of the pleasure to be derived from our trip across the mountains and we heard many prophecies in regard to our going to destruction. In fact the last words we heard were "God bless you old fellows – good-bye; this is the last time we will see you," &c., &c., not a very pleasant starter but we came off in no very desponding frame of mind. We found the trail for a short distance very good, it having been kept broken by some ranch men who live a short distance above, but it was hilly and side hill at that, so that with upsets, broken sled and other disasters being the results, we found ourselves at dark only

3 miles from Quesnelle. We struck for the last house and got to Pollock's[18] at 6, rather used up and having left one load behind. Pollock was kind and gave us a supper and a place for the dogs who also were played out.

The next day we mended broken sleds and broke a track a few miles out so that the start might be a good one. I also went back and brought up the cached sled. On the 10th we made a fresh start, and left some of our stores at Pollock's, as the sleds were altogether too heavy for such a trail as this promised to turn out, and here I might say a word about the trail. It was built by some telegraph company (I forget which)[19] who proposed to run a telegraph line up north to Behring's Straits and thence to Asia by a short cable. The line was actually in working order for some 200 miles, when the news of the success of the Atlantic cable put a stop to the operations. The director and promoters of the scheme died of the disappointment, and the company left everything as it then was. The wire now hangs broken and twisted from the posts, the greater part of the offices are burned down and the only result of such a vast expenditure of money is the trail we take on our way to Fort George. On the 10th we took a final leave of civilization and started off. After a hard day we made camp only 7 miles from Pollock's or 10 miles from Quesnelle. Our camp was most primitive, being a piece of cotton thrown over poles stuck in the snow sloping towards the fire. This served to keep the wind from our heads at any rate and we certainly were able to enjoy a good sleep after the day's labours.

On the 13th we were 45 miles from Quesnelle, having had some fearfully bad trail over side hill and deep snow. Of course side hills are good enough for mule trains but when you try dogs you will find they won't work worth a cent. The dogs go straight enough but the sled won't keep after them, being more inclined to seek the valley below. So as you can imagine it requires a good deal of work and patience to keep the sled in the road while the dogs haul.

The 13th was Sunday, and we had a very heavy fall of snow, but were able to make 12 miles that day. As the snow was now very heavy I gave my train to Johnny and went ahead with Jarvis, who in addition to the work of breaking track had been very busy all the time counting his steps,[20] so as to get the correct distance. Henceforth I shared his labour, and I can't say that I like pacing distances. Hard work it is to break track, but when you have anything to think of it is pleasanter. But when you walk all day and think of nothing but 1, 2, 3, &c., &c., it is monotonous enough for anything. However, all this is a part and a necessary one of the proposed exploration, and I shan't growl at anything we have had so far.

We had a hard bit of work at the Blackwater River,[21] 50 miles from Quesnelle. The river is bridged by poles and telegraph wire, but on this side it is bare ground and the hill is very steep indeed. The poor dogs did their best to get up, but the end of the matter was that we hauled the loads and they looked on. I went back to my own train, and with three men hauling, we got it up at last, but I am afraid my whip did more than its share of duty that day. We got on top of the hill about noon, and had rest and lunch there. We now had about two foot of snow, which was very soft and clung to our snowshoes in great masses; it was also very hard on the dogs, this wading through snow, only freshly beaten down by two pair of snowshoes.

On account of the heavy snow we had, on the 16th, to make another cache (No. 2) of provisions, stationery box, &c., and on the 17th one of our sleds rolled down a steep side hill, and when recovered wasn't worth much, except as kindling wood. The dogs were all right; how they manage themselves I don't know, so we had to cache what stuff we could spare, put some on the one remaining sled, and take the rest on our backs, the four dogs running with only their harness to trouble them (Cache No. 3.)

I forgot to mention that after crossing the Blackwater we left the telegraph trail, which goes on north, and took a C.P.R.[22] trail to Fort George. This latter is if possible a worse one than the telegraph trail. At noon on the 18th, as we were at lunch, an Indian from Fort George came along, and in reply to our questions said we wouldn't get to the fort that night, as it was "siah," a long way. This Indian had a small dog, on which he had his kettle, blanket and grub, he himself carrying the axe and some fuel. Happy thought for us, why not make these beasts of ours do some of our work, and take the packs which are wearing our shoulders away. No sooner said than done, we loaded them and started, Jarvis ahead, counting one, two, three, I next, calling along the packed dogs, and Johnny behind, poking up the lagging ones with a stick. Alec drove the sled behind. It was a comic sight to see the dogs who had never packed before, go rolling from side to side with their loads. As sure as one would try to jump a log, the weight of the load would tumble him back, and if he did manage to get on the top of the log, the weight would tumble him forward in the snow, where he would lie till helped up, but they soon got used to it and were able to follow us, and we went at a good pace, being on a hard track and in a hurry. At any rate we got into Fort George about 5 p.m. that (last) night, though Alec and his train didn't arrive for some hours after. Distance by our pacing, 125 miles from Quesnelle. By the river it would have been 83 miles. We spent 12 days on the way, one of which was at Pollock's. Greatest distance we did was on the last day, 23 miles. On the way we had used up one sled completely, and the other is fit for nothing now. We made three caches, containing in all about two-thirds of our original loads. This looks bad for our future journey, of which this is scarcely a beginning, but then the road we have come over is a most fearful one, while the river which we will follow from here will we trust be much better. At any rate, as we express it, "the country is quite safe," meaning *we* are quite safe. The country between Quesnelle and here is wooded, in some places burnt over, in others green. It is very hilly and broken, and the trail generally runs from the top of one hill to the top of the next, making it first rate for a telegraph line, but very tough on the dogs and us. On the whole it is the worst place I ever saw to do this kind of travelling, and I shall never try it again.

We found Fort George in charge of Mr. Bovil,[23] a son of the Chief Justice of England. With him is staying Charlie Ogden[24] from Stewart's Lake Post.[25] He, the latter, came down to help us get a fair start, and seems very ready to put us in the way of getting dogs, men, &c.

After supper last night we lit our pipes, and we spent the evening discussing the plans to be adopted &c., &c. Ogden is pretty well posted in the country. Bovil is just out from England, and consequently very green in these matters. He is a gentleman and a good cook. As his rations in the H.B. Co.[26] don't amount to more than 25 lbs. dried salmon per week, flour and tea in addition, he won't have much chance to exercise his knowledge of the culinary art. At present *he has killed one of his working oxen* and we are living well. What he will do for his next year's crop I don't know, but he hates the sight of a dried salmon and I hardly wonder at it. I'll put some more to this shortly.

FORT GEORGE,
Dec. 20th.

On the 19th we had a square loaf of which we all stood in need. Then having on that day got a new sled and an Indian (Quaw), Alec and Johnny started back to Quesnelle for the caches. They left early this morning, Quaw going part of the way. He will return from cache No. 3 with articles (books, sextant, &c.) which we want here. Ogden leaves tomorrow for Stewart Lake and he will send down some dogs, dried salmon and sleds from there.

Salmon and dogs are scarce articles at Fort George, and as we want them, of course, the noble red man won't sell except at exorbitant prices. This shows that civilization has been making rapid strides among the Indians of British Columbia.

FORT GEORGE,
Dec. 26th, 1874.

MY DEAR EDWARD, –

From the date of my last letter we spent the time reading, smoking and having a very comfortable time generally. Occasionally we took a walk on the river, which is frozen hard and very good traveling. Getting ready for Xmas was a novelty. We helped Bovil to make a pudding, and he seems to understand the business perfectly. Christmas day was very cold indeed, but a very pleasant one nevertheless. We dined at 6 p.m., and I enclose a bill of fare, that you may know that we had grub, if other things were wanting.

Soup, clear, (*à la* Bovil.)
Fish, salmon, (dried *à la sauvage.*)
Pièce de Résistance; roast working ox.
Entrees, turkey (*à la grouse.*)
Vegetables, potatoes.
Plumpudding and brandy sauce, pipes, tobacco and a glass of brandy and water, to absent friends.

Since my last, we have had a few inches of snow, but the weather generally has been fine and very cold.

My dogs are as lively as crickets and are getting as much salmon as they will eat. The trip from Quesnelle galled some shoulders, but they are rapidly getting well under my care. A train dog isn't very loving but these are very fond of me – *at feeding time.*

FORT GEORGE,
January 7, 1875.

MY DEAR EDWARD, –

After 'Xmas we began to look for Alec every day and finally to fear that he had fallen into the river which he had been ordered to follow on his return. Our time was spent in taking long walks up and down the river and in cutting a trail around some open water a few miles above; the season is getting on and it begins to look as if we wouldn't get off before spring. Still we flatter ourselves that the "country is quite safe." Today we started an Indian down the river to look up Alec, giving him orders to bring him dead or alive, so I hope we will hear something in a few days more. We are O.K., the dogs ditto. Bovil has a queer specimen of a cur which he fondly imagines is well bred. His dog's name is Jack, he doesn't know it himself, but Bovil says so. His obedience is really wonderful, when Bovil says "come here Jack," he starts at once to get under the bed, and then his master says, "that's right go and lie down under the bed, you beast," or else he gets the beast in one hand and a dog whip in the other, and makes music for the million. We are having some snowshoes and toboggans (dog sleds) made while we wait; ours are pretty well used up in the Quesnelle trip.

FORT GEORGE,
January 13th, 1875.

MY DEAR EDWARD, –

On the 8th of this month, the day after we started the Indian to look for Alec, he returned having met the youth down the river a day's travel. Alec had a hard trip, having brought the dogs, sleds and loads up the river as far as the 1st canyon in a canoe. He then started up the river with the train, but the ice was very bad and he finally had to leave his load, the dogs being played out for want of grub. You see he had started from Quesnelle with six days' grub expecting to make quick time on the river. At Blackwater he got an Indian to help him and together the three packed a good part of the load up, one important part was a mail containing letters from home. The 8th was very cold, – 47 degrees, but as I had given you a register already, I won't repeat all the thermometer readings in my letters. Alec and Johnny returned with fresh dogs the next day after arriving and brought up the sled, &c, on the 12th; between the 9th and 12th we had sent off a H.B. Co.'s servant to buy salmon for us, and on the 13th he arrived bringing with him a messenger from Ogden saying that the trains would be on hand in a few days. So we are now all ready to start on the Smoky River Exploration, and will leave tomorrow, I think. Before I give you an idea of how our loads, &c., are made up, I must tell you of a fight we had in the house this afternoon. We were sitting smoking quietly when the door opened and in walked an Indian, he made straight for Bovil, and before you could say "Jack," he hit at him with a hardwood club made for the purpose; fortunately Bovil caught part of the blow with his arm or it would have been the last of him, he then jumped up and grabbed the Indian and around the room they waltzed, each trying to get a good blow; at last I saw the Indian feel for his knife, so I took a hand by getting my dog whip and putting the handle into Bovil's hand. The handle is loaded you know, for the purpose of knocking down a refractory dog. Well, as soon as Bovil felt his weapon, he jumped back, broke away from the noble red, and gave it to him *good*. After that we had no trouble in dragging him to the door, where he remained some time after recovering, with the blood running down his face and his knife in his hand, ready to let daylight into our host. There was great excitement among the Indians, who gathered outside in crowds. Finally the savage was coaxed off and I was as glad as any one to see him go, tho' I had a good six shooter and wasn't much afraid. It seems that an Indian boy had told stories, lies, about Bovil and some squaw, for which Bovil kicked him well, hence the row, in which the father sought to revenge the kicking of his son.

All's well that ends well, but Bovil better be careful with these brutes.

Here I will give you an idea of what we have to consider in making up our load:

First, grub per man per day, 4 lbs. 4 men	16 lbs.
〃 dog 〃 2 lbs. 8 dogs	16 lbs.
	32 lbs.
Blankets, instruments, kitchen, &c., &c., for 4 men	150 lbs.
2 dog trains will carry, No. 1	250 lbs.
〃 〃 No. 2	350 lbs.
	600 [*sic*]
Subtract	150
	32)450 (14 days.
	32
	130

So we leave to-morrow morning with only 14 days' grub for all, but at "Hanington's cache"[27] there is bacon and flour and at Bear River[28] we will get some salmon. So when the Stewart Lake trains arrive we will be able to go on again with full loads. The C.P.R. rations amount to 4 lbs. 5 oz. per day, and *it is all eaten*, the air in these mountains giving one a great appetite. I will give you the ration list on a separate sheet.

I have been puzzling over our supplies and from what I can make out we will have some small rations before we get through, but our instruments must go or else we might as well stay. The weights I have given for our dead weight (articles not grub) are under the real weight I am sure.

But Good Night,
Yours,
C.F.H.

CAMP NO. 4, FRASER RIVER,
17th Jan'y., 1875.

MY DEAR EDWARD, –

We got away from Fort George on the 14th Jan'y. about 2 p.m. and camped 7 miles up River. Bovil looked very sad as he won't see any white man till next spring. We took with us "Quaw," an Indian who has fish on Bear River (see plan) and "Te Jon" an Indian boy. It was very cold – 53 degrees and my nose as usual got fits. We camped in the old style with boughs at our backs and under us and a good fire in front. Each man has two pairs of blankets and all his clothes on his back. This avoids loading the dogs too heavily and at any rate one needs them at night. The first night I can't say I slept very peacefully, – 53° don't allow that, but I slept a little and that's something. We were up early next day and had breakfast at once, no time being lost in washing or dressing. I found my leader ("Marquis") with both fore feet frozen hard that morning, so I had to let him run loose and try three dogs. We killed a grouse to day, "Quaw" having a gun with him. I forgot to say that we have a rifle and cartridges, but we left the gun in Quesnelle, too heavy to pack. This morning poor old Marquis' feet were in a dreadful state, having frozen and thawed several times, so he had to be shot and it was done accordingly. Jarvis did the deed and we left the good old brute at our last night's camp more comfortable than he has been since he froze his feet. We had some hard travelling to-day through the Giscome Rapids[29] where open water kept us off the river. All the dogs are lame, very lame. Their feet get wet and the snow sticks to them, then of course the poor brutes pull the lumps off with their teeth and in the hurry they *bite their toes fearfully*, but we can't help that and they must go on sore or not. I can't imagine a quicker way to harden a man's heart than to put him driving dogs.

This is Sunday in civilization, the only thing we have to remind us of the fact is the date in our diaries and I suppose we won't have one till we get to the east side of the mountains. Our camp tonight is 52 miles from Fort George and about 12 miles below mouth of North Fork. Still very cold indeed.

HANINGTON'S CACHE, CAMP NO. 7,
Jany. 20th, '75.

MY DEAR EDWARD, –

Here we are at last, 82 miles from Fort George, in 7 days. We came along pretty well, though as I said before the dogs were very lame indeed, and the travelling bad in many places. We saw a ptarmigan on the 18th, the only one I ever came across, a very pretty white bird, smaller than our partridge, and very fond of snow clad mountains, where it stays in summer. We have four feet of snow, and find it hard work to shovel out room for camp, but so far we have done it always, it has been very cold and my poor nose has caught it often. A common wind in summer would freeze it anyhow I believe.

These dogs of ours are rather used up, but I have shod most of my three with deer skin shoes and they are getting over their lameness. The Chief (Jarvis) and Quaw now generally break track ahead of the trains; about noon they are sometimes half a mile ahead, when they stop to make a fire for lunch. N.B. At the first stroke of the axe, dogs which a moment before could scarcely crawl, prick up their ears and take the load along as if it were nothing. It's no use to yell, "ulwa" they won't stop till they get to the fire. I think it would be a good idea to keep a man ahead to chop the dogs along, instead of having one behind for the same purpose.

We got here at 9 a.m., and I leave in an hour for Salmon Cache, taking Quaw and Te Jon with me, also two trains empty to bring up a fish supply.

While we are away, Jarvis, Alec and Johnny are going ahead to break track and we hope to hear of the Stewart's Lake trains before we meet here on my return. There is a good deal of fresh snow on the ground and the river in consequence is overflowed. This as you can understand makes it lively for the dogs, and gives us exercise in hauling dogs as well as in driving them. But I'll close this for the present as I must leave for Bear River.

C.F.H.

SALMON CACHE, BEAR RIVER,
21st January, 1875.

MY DEAR EDWARD, –

I left camp No. 7 yesterday at 11 and with the light trains over very bad snow made 22 miles to last night's camp. I would have gone further than that but Quaw, who broke track, was used up, so I stopped. This morning we were at the portage early, and started across it, Quaw said it is good for dogs. Quaw is a liar, and I believe he never saw the portage before, at any rate he couldn't find it half the time. However, after much climbing, hauling the dogs up and letting them down perpendicular places by ropes (the truth) we got to Bear River at 4 p.m. Distance across portage about 3 miles, which we did in 7 hours hard travelling, I am mad tonight and have been giving Quaw a piece of my mind. After getting to Bear River, we came up 5 miles and found the cache in a good state of preservation. Quaw has quite a house here and in it we now sit. Brush on the floor, a good fire and a dry roof overhead, all make a very good picture, which I'd like to sketch.

The salmon caches are outside in some large pine trees, where the bears and wolverine cannot get.

HANINGTON CACHE,
26th January, 1875.

MY DEAR EDWARD, –

On the 22nd I entered into a discussion with Quaw as to the benefit to be derived from a cheap sale of salmon. Among other things I told him he would certainly have a fair chance of going to Heaven when he died, all of which being spoken in the chaste language of the Chinook, he took into his heart. The end of it was, that after breakfast he handed over 650 salmon at 10 cents each, and he also helped to pack them on the two dog sleds. The noble red man is a strange individual. Last summer when the salmon were running up the river, and we wanted some, fresh, Quaw wanted us to pay $1.50 each for them, now after having cured and dried them he sells ten for one dollar. Quaw says he has been up the pass we are going to explore and that it is good, but he won't go as guide at any price. I tried him again when at his cache, but no go; he says "in three days journey you will get to a fork of the N. Fork, take the left. In two days more you will strike a fall as high as a tree, which you will have to portage around. In 5 days more you will see meadows and a very small stream running through. After that you will travel 3 days when you will find water running east, and you will see the sun rise out of the prairie." This is a very good prospect for us, if his word can only be depended upon, but I'd rather see the old chap go up as guide than hear all his ways and means of getting through the pass.

The great point is, how long are Quaw's "suns" or days, but that we'll find out in time. Well, as I said before, I got the salmon loaded on the sleds and having given Quaw an order on the H.B. Co. for his money (which they will probably pay in goods at 500 per cent, profit) I said good bye, wishing him success in his trappings, &c., and left. Te Jon takes one train and I my own and we came around by Bear River, the portage being as I said before; the travelling on Bear River was very bad and I soon had to make Te Jon drive both trains while I broke track. We got only five miles down river and camped, the dogs being completely played out. I broke track a few miles ahead after we got comfortable for the night. The next day was a little better and we made 8 miles, getting two miles below the mouth of Bear River, but I had to leave one sled at Bear River, and putting 8 dogs on the other drove to camp. Then while I made ready for the night, Te Jon went back and brought up last sled. In the evening I broke track ahead. The next day we found the river frightful, the water having overflowed on account of the heavy snow. As you can imagine, the sleds stuck fast in this slush, and we have to get poles, turn them (the sleds) over and scrape off the bottoms, then we go on a few more yards, when we repeat. It was fearful both on men and dogs, and I was delighted when I saw the N. Fork[30] on which hoped to see some remains of the track we made on the way down; we got there about 3 o'clock on the 25th, with one sled and 8 dogs hauling, so I set Te Jon at the camp and went back for the remaining sled; 8 dogs make a fine train, I can tell you, but they have had such a hard time of it that their spirits are about broken. It was late when I got that sled to camp and it didn't take much cradling to send me to sleep. This morning I made up my mind to leave one sled here, and with the other and all the dogs, go to my cache before night. This I did and we left camp early. It snowed hard all day but we got here at 4 o'clock, altho' we had some overflowed ice to work with.

I forgot to say that three miles from here we found a fresh track and after that we came in in fine style. I found Jarvis and the others here, having returned today from the trip up river. They report a good track ahead now, but tonight will fill it up I think. A heavy snow storm. So far we have never seen a track remain open more than one day, but this may be an exception to the rule.

I have had a square feed to-night. My stock of grub having given out some time ago, and my taste for salmon not being developed yet, I have been hungry.

Jarvis is sorry that Quaw didn't come back as guide, but as usual we agree in saying that so far "the country's quite safe." But I'll turn in – good night.

C.F.H.

HANINGTON'S CACHE,
January 28th, 1875.

MY DEAR EDWARD, –

On the 27th Alec and Johnny, with two trains of dogs, left to bring up the sled I had cached at the mouth of the river. It snowed nearly all day and camp was most miserable in consequence. This morning we had a consultation and concluded that the Stewart Lake trains would fail to appear, so as we are bound to get through this pass, we set to work to make a toboggan to be drawn by ourselves. We had it in a good state about 6 P.M. when in came Alec and Johnny with trains, and with them three trains from Stewart's Lake. A very agreeable surprise to us, I can tell you, trains loaded with salmon and drawn by good looking dogs. The drivers are Hassiack, Ah Kho, and Tsayass, smart looking fellows. So we are now in good trim and high spirits. We had letters from Bovil, who reports that the pugilistic Indian came to him and asked to be forgiven, so they are now friends and happy. Alec met the trains at the mouth of the river and they made good time up here; I am delighted at the arrival, as I never was fond of hauling a toboggan through four or five feet of snow. Jarvis looks happy and relieved in mind.

I'll turn in now and finish the night thus.

Yours,
C.F.H.

THE FORK, CAMP NO. 9
30th Jan., 1875.

MY DEAR EDWARD, –

We spent the 29th in packing the sleds and getting everything ready for a start. We have now 5 trains of 19 dogs. Tiger was shot on the 29th, as his lameness prevented him from doing anything but eat grub. We had about one month's grub when we came to examine it; that is one month's full rations for men and dogs, but we won't use full rations, so it must last longer. We started bright and early this morning, and found the track, for the most part, drifted full, sometimes it would be visible and then we made good time, the new dogs doing very well.

We had to leave some bacon and beans in my cache, the dogs not being able to take all of it. We did 18 miles to-day, and our camp is at the Forks to-night.[31] If this be the place meant by Quaw when he said 3 days' journey to the Forks, we must have walked very slowly. I believe though that he knows nothing about the country. The North[32] and South[33] branches are here about the same width, 200 feet. Our camp is between the two.

CAMP NO. 12
Feb. 3rd, 1875.

DEAR EDWARD, –

We followed Quaw's instructions, and took the North Branch. On the 31st we had a blinding snowstorm, which filled the track completely and didn't surprise us by doing so. The weather has been pretty cold and the travelling bad now. Took an observation at noon on 1st and made Lat. 54° 26' North. On that day Sam's shoulders were so much galled that the beast couldn't work and was turned out to run; on the 2nd we came to the conclusion that this sort of work will kill the dogs completely, so Jarvis started off with one train and three men, while Alec, Johnny and I spent the day waiting for a track to be made. In this country a track made in the coldest weather will with one night's frost harden so that it will bear dogs and loaded sleds easily; hence the two parties. I enjoyed the rest very much and did some mending on trowsers and shirts, duplicates of which I have none. This morning I had the camp up long before day and we had to wait for light to show us the dogs. We made good time over a capital track, but it wasn't much use, 9 miles up we found Jarvis making a portage around "falls as high as a tree,"[34] so this afternoon we all were at that and got through about 5. The portage is 3 miles long and after passing the falls we have some very bad canyon to go through. The river is open for the most part, and we have only a narrow ledge of ice and snow to make a track on. On the right rises perpendicular rock 400 or 500 feet high, on our left is the river roaring and rushing 20 feet below. This ledge was formed when the water was high and when the river subsided it was left. High water mark is here some 60 feet above our heads and it must be a grand looking place at high water.

To-night we are all in camp here, where we returned from portage building.

We begin to believe that Quaw is not a liar, sure this fall is as "high as a tree" but his days must have been short.

Yours,
C.F.H.

CAMP NO. 15,
8th February, 1875.

MY DEAR EDWARD, –

On the 4th we had a very heavy snowstorm, just to make things lively I suppose. Jarvis left early with two men and a very light sled, while the rest of us spent the day in getting the stuff over the portage and through the canyon. It was a bad trail and the late snow hadn't improved it at all. The first trouble was a steep hill, about 150 feet high, and it took us all to get one sled up at a time. The dogs didn't seem to care for the place at all and when two men would be hauling and two pushing the sled, ten to one the dogs would turn about and go down the hill. Tough on the whip. However, we got everything over safely about 4 P.M. and camped at end of canyon. The next morning we were off bright and early and went only a short distance before we struck another canyon and fall[35] and had to follow another portage made by Jarvis. It wasn't very bad and with 8 dogs on each sled we took the stuff over in fair style. The worst part was at the further end where the descent to the river was almost perpendicular. Here the sled invariably reached bottom before the dogs, though the latter did their level best to get out of the way. It was killing work on the beasts; how they stand it I can't see.

After crossing the portage we had canyon all the afternoon and after working hard, very hard all day, we camped just 6 miles from last camp. I broke track all day, Jarvis' trail having drifted full. I managed it thus: Started off about 5 miles an hour and walked away some distance, then back to the dogs and then forward again, hoping to give them the benefit of three pair of snow shoes. We didn't get to camp till 6 and the dogs couldn't have gone another mile. Poor beggars, sometimes I am sorry for them, but that don't pay, we can't afford to rest them or ourselves and we both need it.

On the 6th we had very warm weather which made it worse. We passed another fall,[36] and had the usual amount of hauling, dog whipping and general hard work. Road bad enough for anything. Overflowed with water and no track visible, though Jarvis and men passed over it only the day before; at 10 A.M. on the 7th we came to the forks (No. 3)[37] and I got a note from Jarvis with instructions to follow N. Fork up to where I'd find his sled. This I did and camped there. Jarvis came in in the evening, having been to head of South Branch[38] and found no pass. He had also fired at a moose but the distance was too great and he didn't hit. A moose would be a great thing for us, as with it we could take a rest and feed ourselves and dogs. Though we are not very hard up yet if the rations be a *leetle small.*

To-day I have been 7 miles up this the N. Branch. Above camp 1 mile is a fall of 200 feet which I went around via side of mountain, above that the river widens out and meanders through muskeg and meadow for three miles, then it gets rapid and canyony and turns to north. Evidently it rises in the mountains and comes from glaciers.

CAMP NO. 15. Continued.

I turned when I had reached an elevation of 4,000 ft., which is 250 ft. higher than the Yellow Head Pass;[39] so that from an engineering point this branch is worthless. There is nothing left for us now but to retrace our steps to Camp No. 9 and try the south branch. This we will do tomorrow. We both feel that it is tough to turn back but the quality of the Smoky River Pass must be known and we are going to find it out. The Stewart Lake Indians are delighted, as they think we are going to Fort George. Johnny is as unmoved as ever, and Alec ready for anything. So goodnight.

Yours sleepily,

C.F.H.

THE FORKS,
13th February, 1875.

MY DEAR EDWARD, –

The night before we left the camp (No. 15) at head of N. Branch of N. Fork of N. Fraser River, we had a heavy snowstorm, and consequently had no track to return on. This was expected, as so far it has always been the case. I won't trouble you with a history of each day's journey over familiar ground, except to say that the portages around the canyons gave us as much trouble and hard work as the up trip. We got here yesterday having done the 68 miles in 4 days. Capital time considering the road we had. It would have been amusing if not so serious, to see the number of times a sled had to be turned up to get the slush off its bottom in each mile. However, we arrived safely and are glad to be here. This morning Jarvis sent Alec, Johnny, and Ah Kho, ahead up S. Branch to make a trail, as 'tis the only way we can work at all. They took a hand sled as the snow is too deep to take dogs without a track. Tsayass and Te Jon were sent to Fort George with seven dogs and one sled. They are to get their provisions at my cache and seem to be delighted at the chance of going away. Before starting they said goodbye to their Indian friends and said that was the last they would ever see of them. This was comforting to them to say the least of it. We sent Sam out by them, also Chun and Tyepaw, a Fort George dog, which I drove as long as he could go, and these with Tsayass' train made up the seven. We have been going over our supplies, and by a free use of the Multiplication Table, and some addition thrown in, we find that we still have *one month's grub on hand.* Just the same as we had two weeks ago, some sense in this kind of grub which gets larger every day. We have invented a scale by which we can weigh our stuff. 25 rifle cartridges make a pound is Alec's belief, so on that we work, with a bag of cartridges on one end of a stick and some grub on the other, we lay out each day's grub for each man, and the beauty of the arrangement is, that we can shorten the rations by taking out a few cartridges while it appears as if we were giving good weight. But it is cheating our own stomachs after all, and I would like one square meal occasionally.

CAMP (17)
16th February, 1875.

MY DEAR EDWARD, –

Early on the 14th we left camp, and went up the South Branch.[40] Imagine our disgust when 3 miles up we found Alec making a portage around a "fall as high as a tree,"[41] we camped and the whole party went at it in the afternoon. The next day Alec left early to finish portage and go on ahead with the trail, while we took over half loads, dropped them at the end and came back to camp. This was the worst portage yet, very full of holes and hills and fallen trees, and dogs take advantage of these places, when they get stuck going up hill, and the driver takes the rope ahead to haul on, the brutes turn about and go; then after pulling and working your hardest you get them to the top, away they go down the other side, and no power on earth can stop them till they get to the bottom, where when you reach it you will probably see sled and dogs piled in a confused heap and well used up, and it takes a good humoured man to get them started again without some swearing and a good deal of whip. I regret to say that I haven't a temper of that kind. The worst place in the trail was where it takes to the river again. Alec made that part of it, and I think he must have let his sled go. My dogs took a run when they got near the top and laughed when I called "ulwa"! I held on to the rope as long as I could and then let go when dogs and sled disappeared. When I got to the edge a pretty sight met my view, about half way down stood a tree, on one side of this was the sled and on the other the dogs in a sweet state. Evidently they had been unable to get out of the way of the sled and tried a side trail with the above result; I cut the tree, and told them to go to _____ below and they went, the sled first and the dogs yelling and struggling after, lastly your humble servant who had some broken harness to mend and some bruised dogs to attend to when he got down.[42] We went back to camp after the one trip and found Sam and Chun who had evidently broken away from their keepers.

This morning Jarvis shot them both after asking me to do so. I couldn't shoot Sam at any rate. He worked himself nearly to death for us, and it is too hard. But we haven't very many salmon for them now and can't afford it. This was a very warm day with snow and drizzling rain, we got over the portage all right and camped in Alec's camp of yesterday, 14 miles from the Forks; I hope we have got clear of those canyons, a few more would kill our dogs completely, and our own condition wouldn't be improved by them. For my own part I can only say that on this trip I have worked harder than ever before; physically I mean, otherwise there is little to do.

We are travelling through an unknown country without a guide and take things as they come. Good night.

CHAS.

CAMP NO. 20, NEAR TURN INTO THE PASS,
19th February, 1875.

MY DEAR EDWARD, –

On the 17th nothing of note occurred; on the 18th we met Alec, who had been to another fork, and didn't know which one to take, as to him they both looked unpromising. We camped at the Forks,[43] and I went up the South Branch, while the chief[44] and Alec tried the north. We found that the river here takes a turn to the north, and that the South Branch is a short glacial stream, though looking at the place from here, you would think the mountains shut the whole place up. To-day Alec went off again and we spent a very jolly evening, having found what certainly seems to be a pass through the mountains. We are now 47 ½ miles from the Forks, and have found this branch so far a great improvement on the north.

CAMP 22, 21st February, 1875.

MY DEAR EDWARD, –

On the 20th we had a first-class trail and did seven miles before noon. Camped in Alec's last camp, and a queer place it is. The valley is here about a mile wide, the river running through meadow and muskeg. Our camp is on an island in the centre, and all around are the mountains, some of them beautiful, if we only had the spirit to enjoy their beauty. Today it snowed all day, and of course the track was full before night, and not to be found. We did 11 miles though, and didn't say much, though we thought a good deal, I fancy. So far we haven't been able to keep an open track more than one day. We have plotted up our work to this Camp No. 22, and find that in a straight line we are only nine miles from Camp 15, on the North Branch.[45] If we had only known, what a lot of time and distance we could have saved, and our dogs would have escaped some 100 miles of travel, poor brutes; but as we didn't know it, it can't be helped.

CAMP 25, 24th February, 1875.
AT THE SUMMIT.

MY DEAR EDWARD, –

On the 22nd we met Alec coming back as he didn't see the use of breaking track only to have it filled up by snowstorms, so we went 7 ½ miles and camped with him, after that I broke track a few miles ahead.

On the 23rd we came to falls[46] and canyon after canyon, and had a good deal of trouble in hauling along dogs and sleds too. I saw Jarvis stop once and begin to think over the situation, so I stopped, too, in fear and trembling, for I was afraid he would give it up. Presently he came along and said: "Frank, do you know what I was thinking of?" I said, "Yep; don't go back for God's sake." Well he said that if we all come to grief he would be responsible, and it was a bad look out now. But I told him I'd be responsible for myself, Alec didn't care about going back, and as for the Indians if they starved or not it didn't matter. So on we went to my great delight, for I'd sooner be found in the mountains than give up the ship. Though, so far as a railway is concerned, this pass is of no use.

After camping at the foot of a fall,[47] the two of us walked on and climbed a mountain to 5,500 feet.[48] Here we saw an apparent fall to the east and our hearts beat high, so we returned to camp and said nothing.

To-day we made a portage and started the men getting the loads up the 2 miles, while we went forward to explore. We found the summit, think of it, at last. This branch flows out of a chain of 5 lakes[49] which lie 5,300 feet above the sea,[50] then you cross a short muskeg containing a lake which flows nowhere, then a little more muskeg and a lake out of which trickles a tiny stream *running to the east.* We went down this stream about a mile to be certain, and then we took a drink of the blessed water, which was the sweetest thing I have drunk for many a day. A splendid view from summit. There are no high mountains in the far distance except one peak (Smoky Peak).[51] It looks like a park inclining gently towards the east, studded with oak, and carpeted with grass (it would be if the snow were not so deep).

Smoky Peak resembles Mount Ida.[52] One rises in striking grandeur to guard the western side of the pass, while the other guards the east. They both present the same aspect, solitary, with their white summits in the clouds, glaciers covering their sides to the line of vegetation, and then the blue and green of the forest covering, they are indeed grand sights and worthy of an artist's brush.

After the discovery of the water flowing east, we returned to camp in high spirits, hurrah, had a drink of Brandy – hurrah, and had the pleasure of seeing the others as excited as ourselves. It was indeed a merry evening and one I won't forget in a hurry. The country is quite safe now sure. There are 6 feet of snow at this camp and we have shovelled out camp to the ground. So our view is limited when we camp.

CAMP NO. 26,
25th Feby., 1875.

MY DEAR EDWARD, –

We left camp early this morning and made good time across the lakes. At the summit we stopped, marked a tree, "Summit between B. Columbia and the N.W. Territory,"[53] date and names, then with one leg on each side of the line drawn on the snow, we drank the last of our brandy and gave three cheers. I repeat them. Hurrah! Then we started down the creek along which the snowshoeing was very bad. At noon we saw a lot of prairie chickens,[54] but having no gun didn't get any. We did 13 ¼ miles to-day and the creek which was about three inches wide at the start is now about 15 feet.[55] A good fall in it all the way. We haven't "seen the sun rise out of the grass" yet, but we hope to soon. We are as happy and contented tonight as if we had had a good dinner, a thing we have almost forgotten. The camp isn't very blue generally, but there are some puns and jokes going to-night.

But now to sleep.

Yours,
C.F.H.

CAMP NO. 30, SMOKY RIVER,[56]
March 1st, 1875.

MY DEAR EDWARD, –

On the 26th we pushed along as usual over very bad travelling, the snow was hard enough to bear us on snowshoes, but the dogs would go clear to the bottom and stick fast, so we had to break the crust down every step, which was as you can imagine very hard work. Early that morning we were stopped by a fall, the finest one we had yet seen.[57] I crawled to the edge on my stomach, the ice being thin and looked over. The river lay 250 feet below and the trees, &c., looked very small at that distance. On each side of the river the rock rose nearly perpendicularly and altogether it was a hard looking place to get around.

After satisfying ourselves with the view we turned to go back when one of the men proposed a drink. To get it he took a small axe which at the first blow (a very light one) went straight through. You may bet we got out of that in a hurry. We went back about a mile and took to the side of the mountain which we followed with much trouble till we got a mile below the falls. To get the sleds down to the river required no trouble; to get them down whole took a good deal, as it was as near perpendicular as could be. Finally we took off the dogs, turned the sleds on their sides and got down in that way; you can imagine it was steep when I tell you that one sled having got stuck halfway between some trees, I tried to go up to help the driver and couldn't possibly do it, though I did my best. That night we camped late, having done 5 ½ miles and found only one foot of snow in the woods. This will appear strange to you, but the same peculiarity extends along the eastern slope of the mountains for a belt of about 60 miles wide. Beyond the snow gets deeper again.

On the 27th we passed a 20-foot fall,[58] around which we made a portage without much trouble. Just below this fall we struck good travelling hard crust and we did 14 miles that day, passed a branch coming in from south,[59] which Jarvis explored for some distance up.

On the 28th we had a good deal of open water and had to take to land frequently. Another dog dropped today. Jarvis had to follow behind slowly as he is suffering from *mal de raquette*.[60] He doesn't say much about it but when he takes to the broken track with a white face and set lips you may guess he is in pain. I have been doing the track breaking since he fell to the rear, and I begin to feel a little pain in my ankles to-day. Today we did 13 miles and camped early to mend snowshoes which are very much used up. We have got over the good snow and are now in bad travelling again. Snow hard enough to bear us but which the dogs broke up. If we were certain what river this is it would be more satisfactory. At present when it turns to the east, we think it falls into the Athabasca and our hearts beat high. Then we come to a turn toward the north and we are sure it is Smoky River, and must lead to Peace River and our spirits go to zero at once.

Passed another branch from south today.[61]

C.F.H.

86 MILES FROM SUMMIT,
CAMP 33, SMOKY RIVER,
March 5th, 1875.

DEAR EDWARD, –

The 2nd was very unpleasant, a heavy snow storm, river overflowed and deep snow, river turning more to the north and dogs getting awfully used up.

Our camp on the 3rd was at mouth of small stream from south and we had not a very comfortable evening. Another dog died that day, died of starvation, and worse still the river turned more to the north, and that as I said before means Smoky River.

Yesterday we camped 1 mile above a small fork from south,[62] and after plotting up our work concluded that this river is Smoky River beyond a doubt. We saw an old track of a snowshoe, but the maker may be hundreds of miles away by this time. I got very bad with *mal de raquette* yesterday and cannot recommend it as a travelling companion to any one who has to travel every day and all day.

Today we have been in camp all day making packs and a cache in which we will leave our heavy stuff.[63] By observation at noon we find we are in latitude 54° 23'N. We will strike across country from here steering by the compass with our dogs following us. Our packs won't be heavy, very; mine is about 35 lbs., but with *mal de raquette* it will be heavier a good deal. We are going to take one sled, but it will be light, the others remain here "to be called for" I hope, but not by us. We will leave no grub of course, but our sextant, stationery, books, &c, &c., with extra clothing remain here; the last clause doesn't trouble me as I put on my trowsers, drawers and shirts at Quesnelle and won't take them off till we reach Edmonton.

By the way, I forgot to mention that at Camp 15, and also at the Summit, we washed our faces and hands. It's a fact. The first time at Camp 15, and then again at the "Summit." I don't know the reason for the first wash, unless it was disgust at having to turn back. The last was a wish to leave all the British Columbia dust behind us. To proceed. I may mention that the men from Stewart's Lake are getting longer faces every day, and they evidently don't think much of this trip, either past or future. Alec is all right and Johnny as good-natured as ever. His constant sentence is "Cultus kopa-jnika. Cultus kopa mika" – "What's bad for me is bad for you."

And now I'll stop for tonight.

Yours,
C.F.H.

CAMP NO. 34,
6th March, 1875.

MY DEAR EDWARD, –

To-day we started early with our packs on our backs, on small rations. Climbed all day, and were glad enough to stop tonight, having done seven miles. We are on a high piece of ground to-night, and before us lies a large valley,[64] so we will have downhill work tomorrow. My pack tonight weighed 300 lbs. at least, and my legs are as sore as Jarvis'.

Yours,
C.F.H.

CAMP NO. 39, March 11th, 1875.
Don't exactly know where.

MY DEAR EDWARD, –

On the 7th we had it down hill till we reached the valley before mentioned. Down-hill travelling is worse for *mal de raquette* than up-hill, though I didn't think so when we were climbing. At the bottom we found a large river, which we thought was the Athabasca.[65] We followed it up a short distance, and then turned off on our old course, following up a tributary which seemed to come from that direction.[66] We turned off for this reason. If this be the

Athabasca, we will, in a few miles, strike the McLeod,[67] and will then know where we are and be able to make St. Anne's[68] easily. If not, it is useless and worse to follow up an unknown river. So we called it "This River." Jarvis and I still kept the lead, though the pain we felt at every step cannot be expressed in words.

As we turned a corner suddenly on the 8th, I in front saw two moose in the river about 150 feet from us. As is usual in such cases, the rifle was in the sled behind, and before we could get it the moose were away and lost. These were the first live things we had seen since we crossed the Summit, and our disappointment was very great when we missed killing one of them, we all stand so much in need of meat.

On the 9th we left the creek, which was as crooked as a cork-screw, and struck across country over valleys, hills and deep snow. Our camp that night was on a creek running north, and probably into the last river we saw. Our meat was nearly finished that night, and our stomachs felt empty.

Yesterday we had as usual very heavy walking across these valleys. In the afternoon when we were on a summit, before us we saw an immense valley, about 2 miles wide. You may imagine our delight: here was the end of our troubles and our want of grub in particular. So with renewed vigour we posted down. When we reached the bottom we found a muskeg with a small creek running through the middle of it, down went our spirits again. Today we came on a creek running almost east so we followed it, and found the trail blazed as if by white men. Tonight everything was jolly as can be though our meat is done and our tea so small it can't be counted. Alec shot a rabbit today, quite a feed for 6 men. River here 60 ft. wide, with grassy banks, etc.

But I must sleep; good night.

C.F.H.

CAMP NO. 43,
15 March, 1875.
Where?

MY DEAR EDWARD, –

On the 12th we had snow all day, and very bad snowshoeing. River 100 ft. wide and running north like the others, we commenced to think that times were hard when we began to eat dog to keep our strength up. Dog too which had been starved and worked nearly to death. I don't believe dog soup is good, but it goes very well. On the 13th we left the river and struck out on the old course about S.E. The Indians from Stewart's Lake went on with their wail about never seeing their friends again. They gave up all hope, and I scarcely wonder at it; still they needn't howl so about such a small thing. Others have friends and just as strong feelings for them, and they may think a good deal, but they don't cry. Yesterday the 14th it snowed all day and we weren't able to see anything. In the afternoon after crossing a river, we came upon a pile of horse dung. It was the prettiest I ever saw and I'd like a picture of that very pile; we examined it and cheered lustily thinking that we must be near somewhere. Buster, my favourite dog, died yesterday. To-day the snow stopped and we saw about 20 miles away a high rock which looked like a photograph we once saw of Roche à Miette[69] at Jasper House. So we turned toward it at once. Tonight we are in camp on a ridge or summit. Before us is a valley, a small insignificant one, which in my opinion contains another creek. Beyond it are some hills and further in the distance a ridge of mountains. So the thing has come down to this: – If the Athabasca be not in that valley it is beyond those mountains. In this case as Jarvis says to me, we neither have enough grub or enough strength to carry us across. So our end will be near here.

You must imagine our camp then to-night. Opposite sit the Indians, Johnny as usual silent and impassive, the other two with their heads in their hands sobbing out their grief as usual too. On my right is my worthy chief Jarvis, very thin, very white, and very much subdued. He is thinking of a good many things I suppose like the rest of us. On my left is Alec chewing tobacco and looking about used up. He had seen "Roche à Miette" once from the east side, but isn't sure whether this is it or not, so he is blue. In the centre I sit, my looks I can't describe and my feeling scarcely. I don't believe the Athabasca is in that valley. I do believe that we have not many more days to live. I have been thinking of "the dearest spot on earth to me," of our Mother and Father, of all my brothers and sisters and friends. Of the happy days at home, of all the good deeds I have left undone and all the bad ones committed. I wonder if ever our bones will be discovered, when and by whom, if our friends will mourn long for us, or do as is often done, forget us as soon as possible. In short I have been looking death in the face, and had come to the conclusion that C.F.H. has been a hard case, and would like to live a while longer to make up for it.

But I am glad since we started that we didn't go back, though this has been a very tough trip and this evening is the toughest part of it.

But I must say good-night.

C.F.H.

CAMP NO. 44, FIDDLE RIVER DEPOT,
March 17th, 1875.

MY DEAR EDWARD, –

The day after that terrible evening of doubt and uncertainty, we went only 6 miles when we struck Lac Brulé.[70] You can imagine our feelings without my trying to describe them. Then 8 miles up the lake to the Depot[71] where we found a family of Indians who set out a lot of boiled rabbits when they found we were hungry. We went for that rabbit and then interviewed the natives. There is no one at Jasper House.[72] This is a disappointment as we hoped to get dog trains and men there to take on east. But the Indians say they can give us some dried deer meat and a piece of mountain sheep. We are all looking very much pulled down; all our dogs are gone but three, and they are all bones and skin. Our one sled is here, and here it will remain. Our distance from Smoky River is 119 miles from summit, 205 from Fort George; we have travelled about

600 miles. The Indians say the track made on Smoky River was by one of their number who was hunting there early in the winter. That the river we followed from summit was Smoky River and also the 2nd one another branch of the same.[73] We are getting well used here. Rabbit straight three times a day. To-night we have our supplies in. Some dried meat and mutton and we start to-morrow. By the map Lake St. Ann's is about 200 miles from here by the way we go. The men want to stay here and go back to Stewart's Lake in the spring. Upon my word I'd like to stay, too; I dread this part of the trip more than anything, although now we have the satisfaction of "knowing where we are." Alec has been over this part of the trip, having come from Red River a couple of years ago. "Roche à Miette" is here all right and I won't mistake it again, should I ever have the honour of seeing it.

The great peculiarity about it is its west side. It is as perpendicular as the side of a house and as difficult to climb. A man by the name "Miette" got up the east side and on top, and it has borne his name since.[74] Rightly enough too.

Well now I'll conclude, very thankful I am that we are thus far on our journey and have been kept through such trial and danger.

Yours,
CHAS. F. HANINGTON.

CAMP NO. 51. MCLEOD'S RIVER,
24th March, 1875.

MY DEAR EDWARD, –

We left the Depot very early, in a gale of wind which blew down the lake, our dogs, Cabree of my train, Captain of Alec's and Musqua from Stewart's Lake didn't offer to follow us as they preferred grub to starvation I suppose; we went down the Lake in a hurry, rather too fast for our own comfort sometimes, and then followed the Athabasca having done 14 miles when we camped, had a little dried meat and a little bread for supper, turned in tired enough. Next day we followed the river 14 miles and then left it to take trail across to McLeod River camp 1 mile on trail. Found the walking warm that day and the rations very small for such hard work. On the 20th the walking was very bad and we only did 8 ½ miles passing a lake in P.M. On Sunday the 21st we did 4 miles to the McLeod and 8 down it on a trail made by one of the C.P.R. parties two years before. Very heavy travelling but the trail is better than the river which was overflowed.

22nd. Travelled 15 miles, 5 to portage across bend of river and 10 to camp. Nothing eventful my diary says, sick of this work, "hard work and deuced small grub." On the 23rd we did 16 ½ miles, 1 to end of portage and 15 more down the river by trail, met Adam, a man from Edmonton, *en route* to Jasper House. He gave us some tea but his grub was about gone as he had been detained by the heavy travelling.

We had a cup of strong tea immediately, and it made us drunk, think of it, drunk on tea. He gave us some sugar which we ate up at once, like Indians exactly and then we pushed on. On the 24th we made good time on Adam's track and did 22 ¼ miles, though a little fresh snow fell in the evening, that is tonight. We start early and stop to rest every hour being not so strong as we once were. At night we stop, Jarvis and I clear up a place for camp, Alec and Johnny get brush and the others cut wood, as soon as camp is made Johnny cooks supper (so called) but long before that I am asleep. I am waked to eat my share, which is measured by the chief carefully and is hardly perceptible sometimes. Then I light my pipe and am asleep before I get a dozen pulls, so you will believe me when I say that I am about used up. Tobacco is the main stay, I chew it all day and smoke in the evening and it is a great improvement on nothing. Our tea now is everything for us, though that first very strong cup made a hole in the supply. However, we boil it over and over very carefully, Jarvis carrying the sack and putting in a fresh grain every time. But we know where we are perfectly and we would have no trouble in getting through were we not so much used up when we left the Depot.

CAMP NO. 54,
27th March, 1875.

MY DEAR BROTHER, –

On the 25th we did 25 miles on the river, our grub getting very short, and the tea nearly gone. I had a sort of fainting fit that day so Jarvis went on a little further made camp, leaving Alec to see me through; all right in evening, plenty of tobacco. On the 26th we followed the river 4 miles and then struck off easterly doing 9 more; had two hares for supper and the last of the bread. The men eat the insides without cleaning them, after they had taken their share of the meat.

Today it has been very warm and hard snowshoeing, we did 11 miles and struck a creek running east. Killed four hares to-day and had a first rate supper; though it might have been better.

I would give anything tonight for a good square meal of bacon, beans and bread, to say nothing of such a one as I often dream of. Still it is well I have something. Good night.

Yours hungrily,
C.F.H.

CAMP 57, LAKE ST. ANN'S,
31st March, 1875.

MY DEAR BROTHER, –

On the 28th we did 18 miles and got nothing. My diary says, very hungry and it says truth. We crossed Dirt Lake or Chip Lake,[75] and camped on the creek, snowing all night. On the 29th we did 23 miles, 15 ½ to Pembina River,[76] 3 ½ down it to Portage and 4 to camp beside a lake.[77] We lived on tobacco and water, and though very weak made very good time with frequent rests. On the 30th, that is yesterday, it was warm and hard walking. Alec lay down several times, but toward evening we met an Indian who acted as guide, so we strained every nerve, C.F.H. in front, Jarvis next, then Alec and lastly the Indians; and we got to this Post at 7 P.M. in spite of all the hunger, weakness and misery. I could have gone a good deal farther that day, with that Indian in front, but when he stopped of course I was played out at once.

Mr. McGillvray,[78] God bless him, set out a supper of white fish, potatoes, milk, bread, sugar and tea and asked us to go at it. There wasn't a word said for about half an hour, and then we weren't able to speak much. For myself I staggered to a lounge where I suffered from the grub as much as I had from the want of it. This morning we were up at 5 and no one being awake I stole some bread. At 7 we had breakfast, a repetition of last night. After breakfast Alec and I took McGillvray's horse and cutter and went to the village, where is a R.C. Mission [see Figure 18], to buy eggs, butter, &c., at one of the half breed's houses (a little mixed this) they asked us to eat and set out grilled buffalo bones, potatoes, tea, &c., and we had a capital feed, at another they gave us bread and milk which we did justice to. Then with a lot of eggs and cream (no butter) we returned to the house and spent the time till noon eating cream and sugar with our bread. At noon another fill, ate all the afternoon and evening and are now as hungry as ever though suffering from the effects of gluttony. This is the end of our great exploration so far as hunger and great danger are concerned. When I think of it, I wonder how we ever got through, for without any guide and knowledge of the country we could hardly expect it. To use Jarvis' words, "It is altogether too large a country for 6 men." My weight here is 125 ¼ lbs. but I'll make up for it in a short time; I have necessarily written this sketch of the trip more in reference to myself than my chief. So here I want to say that the credit of the success of the exploration is due to E.W. Jarvis, whose judgment, energy and pluck brought us through. I only seconded him and did my best. But a divine Providence watched over us all through and we owe him our most heartfelt thanks.

Tomorrow we start for Edmonton, in two sleds with horses, and any amount of grub on board.

I'll write you from Fort Garry[79] if not before.

I remain.
Your loving brother,
CHAS.

FORT GARRY, MANITOBA,
22nd May, 1875.

MY DEAR EDWARD, –

We were two days from St. Ann's to Edmonton;[80] spent five days there; had a good deal of vomiting and diarrhoea, which lasted nearly to Fort Pitt.[81] We were four days to Victoria,[82] and rested there two days. Nine days more to Pitt where we stayed one day. Nine days from Pitt to Carlton[83] where we rested five days. We left Carlton on the 8th May, and were at Ellice[84] on the 15th; stayed there only half a day and reached Portage la Prairie[85] on the morning of the 20th. Then Jarvis and I took the stage and got here yesterday. We left the Stewart Lake Indians at Edmonton to return in the spring. Johnny, Alec, and a guide (Norris),[86] a trader, came through with us. I cannot here give you a description of our day's journey, of securing rides on horseback, riding on carts, camp, and all the rest, but it was jolly, fine weather and plenty of grub. I now weigh 163 pounds, more than I ever weighed, and I feel like a bird, but *hungry* yet. We left Edmonton with horse sleds (toboggans), at Victoria we packed our horses and left sleds. At Pitt we left pack saddles, got some fresh horses and carts and came to Carlton. At Carlton more fresh horses; at Ellice more fresh horses and a waggon [*sic*] which we engaged to the Portage. From St. Ann's to Fort Garry we were just fifty-one days, thirty-seven of which were spent in travelling, and the others in loafing.

Some time I may give you an account of the trip from Edmonton (nearly 900 miles) of the game on the prairie and the prairies themselves. But now I'll conclude by thanking the officers of the Hudson Bay Company for their generosity and good nature. Every one of them did his best to make us comfortable, took us to his own house, though we were perfect strangers to all of them west of Ellice. There I found two old friends of mine, who had partaken of our hospitality when we kept house in Fort Garry, two years ago.

This has been a hard trip from first to last. One that I will never forgot, and never repeat, I hope. I am now in the office here, waiting for orders to go somewhere to work, and making the plan, a tracing of which I send you.

My eye has just caught this sentence in Jarvis' report, which I have been reading (his report to the Chief Engineer, Mr. Fleming): "I cannot refrain from mentioning in terms of the highest praise, my assistant, Mr. Hanington, to whose pluck and endurance the success of the exploration is so largely due."

I put this in because I am proud of it, and I will add that that one sentence from Jarvis is pay enough for all I did through the winter. Jarvis has gone to St. Paul to see a friend, so I am alone here, except that I have any amount of friends who are kind as ever.

And now good bye.

Your loving brother,
CHAS.

The country between Quesnelle and Lake St. Ann's is heavily wooded for the most part. West of the mountains it is much broken and rugged. The streams there are rapid and their banks rocky.

After crossing the mountains things appear in a more settled form, and on a smaller scale, the hills particularly. On the Smoky River there is some fine sandstone, about all we saw on the trip. We had heard of a great canyon on the Athabasca but when we reached it we found the rocky sides to be about 20 feet high and flat on the top. We were disappointed. On the Pembina River there are some coal beds which have been burning for many years. We could smell the smoke about a mile off, and it put us in mind of a city. At one place where the smoke comes out of the side of a perpendicular rock it is particularly striking. The surface of the ground is very hot in many places, hot enough to boil the kettle; and by the way I might mention here that the proper name for "Smoky River" is "Smoking" River, so given from some burning coal beds about 50 miles below where we left it.

The wood about St. Ann's is small and mostly cotton wood. From that to east it exists only in patches and is *very* small. East of Edmonton a man knowing the country can generally find enough wood to make a fire, but there are places where wood has to be carried in the carts, such as the Salt Plain, Pheasant Plain, and some others. I am speaking of the trail which we followed from

Edmonton. There is a trail south of the Saskatchewan where the kettle most of the time has to be boiled by the use of buffalo chips. One reason why we did not take it was the season not being far enough advanced and fires being necessary there.

The map, 25 miles to an inch, is a tracing I compiled from my poor data. It will do to give you an idea of the locality we are now in as well as that followed last winter ('75). It is not correct as regards distance. The trail I have dotted in red, as well as the other part of our journey. The line of the C.P.R. I laid down as near as I could from information gathered in letters, &c. "The plan of our Smoky River exploration" is as correct as can be, having been plotted from the original notes. The camps are marked in red and the elevations in blue [see Figure 15].

You will see that I haven't wasted much time on them, but I trust they are plain and will serve their purpose.

I send the "Smoky River plan" *in toto*, but a very small strip of the other one. Cause, not very much time to spare just now.

Hoping you'll excuse all the deficiencies which can't be helped,

I remain,
Yours sincerely,
C. F. HANINGTON.

DISTANCE TRAVELLED ON SMOKY RIVER EXPLORATION.

	Miles.
From Quesnelle Mouth to Fort George	125
From Fort George to mouth of North Fork	64
From North Fork to Hanington's Cache	18
From Hanington's Cache to Salmon Cache (Bear River), going by Portage and returning via Bear and Fraser Rivers	78½
From Hanington's Cache to the Forks	18
From The Forks to head of North Branch	63
Return to the Forks	63
From Forks to turn into Pass	48½
Turn into Pass to Summit of Mountains	40¾
From Summit to Cache on Smoky River	86
From Cache on Smoky River to *Next River*	9¼
From Next river to Fiddle River Depot	110
From Fiddle River Depot to Lake St. Ann's	217
From Lake St. Ann's to Edmonton	60
Miles	996 [*sic*]

	Miles.
Distances measured by Pacing between Quesnelle Mouth and Edmonton on Exploration	826
Number of paces counted, taking inside figures	2,188,900

DISTANCES FROM EDMONTON EAST.

	Miles.
From Edmonton to Fort Victoria	80
From Victoria to Fort Pitt	113
From Pitt to Fort Carlton	167
From Carlton to Fort Ellice	307
From Ellice to Fort Garry	220
Total	887
	996 [*sic*]
Total distance travelled	1,883
No. of camps between Quesnelle Mouth and Fort Garry	106

RATION LIST, CANADIAN PACIFIC RAILWAY SURVEY, BRITISH COLUMBIA.

For one man, per day and per month: –

	Per day.	Per month.
Bacon and hams	1½ lb.	45 lbs.
Four	1¼ lb.	37½ lbs.
Beans and peas	12½ ozs	25 lbs.
Oatmeal	1½ ozs	2 lbs.
Dried apples and plums	4 ozs.	5 lbs.
Tea	1¼ ozs.	2 lbs.
Coffee	1½ ozs	3 lbs.
Sugar	2½ ozs.	4 lbs.
Rice	2 ozs.	4 lbs.
Molasses		1 gall.
Yeast powder	3 tins to 50 lbs. of flour	
Salt		½ lb.
Mustard		⅛ lb.
Pepper		⅓ lb.
Pickles	Plenty.	
Soap	do [ditto]	4 lbs.
Candles	do in officers' mess.	
Vinegar	do	
Lime juice	do	
Matches	do	

NOTE. –When fresh beef is used instead of bacon, 60 lbs. must be allowed per month.

These rations are used regularly in B.C. The sugar particularly is very often short.

C.F. HANINGTON.

TÊTE JAUNE CACHE, ROCKY MOUNTAINS,
May 4th, 1876.

MY DEAR EDWARD, –

I left Fort Garry in June last, after having spent three weeks very pleasantly there. Johnny thought Winnipeg a fine place; it was his first appearance in a town. He had some money when he arrived, and the first time he appeared after, he was dressed in black broadcloth, swell hat and patent leather boots. His board was paid all the time till a chance should occur of sending him home to British Columbia. The last time I saw him he was sitting by the side of a dry goods store with one arm around the neck of a very good-looking squaw, who evidently thought him no end of a swell. Late last fall he came over here, having been forwarded by express. He had learned to talk English, and when I said, "Iketa mika tumtum kopa okook cola inate la monte?" ("What do you think of the trip across the mountains?") he replied, "Damn hard." He had a good summer of it; lived with the object of his affections (though she did not speak his language nor he hers), and was loud in the praises of Winnipeg. I enquired about the health of his wife, and he informed me that she cried a good deal when he came away; also that should nothing *occur to prevent he would be a father shortly.* So I gave him some clothes, &c., and he started off to his home in the lower Fraser. Alec was hired as a mail carrier between Garry and Edmonton till the autumn, when he was sent to Henry House,[87] 64 miles from here, to look after the supplies there in depot. He came over to see me this winter, and was here on the anniversary of the day on which we reached Lake St. Ann's. We did our best to celebrate it in rum and water, a thing we weren't able to do last year. I had him in our mess, and enjoyed his visit very much, living over past scenes more pleasantly than was possible at the time they were enacted. He went back to his post, and will go to Fort Garry early in the spring. The chief (Jarvis) went to St. Paul, and was sent for from Ottawa, where he was wanted to take a party to British Columbia. Being as fond of this country as I am, and being able to afford himself a rest, he refused and left the C.P.R for a season. After making a visit to P.E. Island and other parts of the Dominion, he returned to Fort Garry and went into the lumber business, where he is now making a good deal of money. I hear from him often. He said once that the mention of Smoky River made him shudder, and I dare say it would. I came out here last summer, and we commenced locating the line from the summit of Yellow Head Pass down the Fraser River to meet another party commencing at Fort George. In November we went into quarters here, and have spent a most miserable winter, the last I will ever spend in this way. We will be at work long before you get this, pushing steadily towards Fort George and civilisation after. The sketch I send of our trip is, I think, full of errors, though not any serious ones. I have written it very hurriedly at different times, with all the din and noise of my friends in arms sounding in my ears.

I know you will make every allowance for my mistakes, which I cannot correct as I've no time to read the whole again.

I know it is written in a rambling desultory sort of fashion, but you'll believe me when I say that I did the best I could under the circumstances. And now I'll say good-bye.

I remain.
Your loving brother,
CHAS. F. HANINGTON.

MINIMUM TEMPERATURE FROM 1ST JANUARY TO 6TH APRIL, 1875, REGISTERED BY MR JARVIS IN THE COURSE OF HIS JOURNEY ACROSS THE ROCKY MOUNTAINS.

Date	Temp.	Date	Temp.	Date	Temp.	Date	Temp.
January 1	-22	January 25	14	February 18	32	March 14	-32
January 2	-26	January 26	-2	February 19	28	March 15	-30
January 3	-40	January 27	-23	February 20	25	March 16	-23
January 4	-10	January 28	-29	February 21	29	March 17	-20
January 5	-38	January 29	-10	February 22	25	March 18	-8
January 6	-28	January 30	14	February 23	11	March 19	-12
January 7	-36	January 31	-2	February 24	-2	March 20	9
January 8	-47	February 1	-23	February 25	-15	March 21	6
January 9	-25	February 2	-29	February 26	-2	March 22	-3
January 10	-45	February 3	-10	February 27	-10	March 23	-12
January 11	-31	February 4	7	February 28	8	March 24	-9
January 12	-40	February 5	-8	March 1	-6	March 25	4
January 13	-50	February 6	2	March 2	12	March 26	-6
January 14	-53	February 7	4	March 3	-11	March 27	5
January 15	-48	February 8	8	March 4	-8	March 28	-4
January 16	-36	February 9	24	March 5	-15	March 29	15
January 17	-41	February 10	-22	March 6	-5	March 30	5
January 18	-45	February 11	8	March 7	22	March 31	9
January 19	-45	February 12	12	March 8	15	April 1	23
January 20	-31	February 13	2	March 9	5	April 2	24
January 21	3	February 14	3	March 10	26	April 3	4
January 22	7	February 15	25	March 11	27	April 4	-4
January 23	-10	February 16	15	March 12	15	April 5	6
January 24	8	February 17	27	March 13	-2	April 6	8

CHAPTER 3

Excerpts from E.W. Jarvis's Diaries, 1875

Jarvis may have kept diaries throughout his life.[1] Seven have survived, for various periods between 1863, when he was a teenager on a European tour, and 1890, when he was a North West Mounted Police superintendent in Maple Creek, Saskatchewan. Fortunately, his handwritten diary of the Smoky River expedition has been preserved, including his sketches. Compiled day by day, when the outcome was unknown, it provides an unvarnished and immediate account of the challenges they faced and overcame.

January 1st – At Fort George – Waiting for dogs & salmon from Stuart's Lake, which are due here about the 10th. River not quite frozen up yet – rim thin.

9th – Alec returned from Quesnelle last night – had a hard time in the river owing to open water and overflowing. He started back at once for part of his load left below Ft. George Canyon.

12th – Alec arrived with balance of supplies. Murdock arrived from Stuart's Lake, 7 days out. The men with sleds are 4 or 5 days behind, and road very bad – he is to go back with Seymour's train tomorrow to relieve them.

13th – Loading sleds, fixing harness &c for a start tomorrow. Murdock refuses to go back at once to the other men – discharged him on the spot.

14th – Arranged a/cs with Bovill and left the Fort with Hanington, Alec, Petit Jean & Johnny – three trains of dogs. "Quaw" goes with us to give 600 salmon from his cache at Bear River. Made 7 m. and camped.

15th – Made 16 m. and camped.

16th – Made 13 m. and camped.

17th – Made 16 m. and camped.

18th – Made 15 m. and camped.

19th – Made 12 m. and camped.

20th – Reached cache at 8.20 a.m. and dumped our loads. Sent 2 sleds back with Hanington Petit Jean & Quaw for salmon. Started up river with Alec & Johnny and one sled to break track – made 8 m. and camped. 82 m. from Fort George.

21st Thursday – Went 12 m. up North Fork and camped – No. 10 – Heavy snow all day and heavy going.

22nd – Snow fell all night – 10 in. in all, and dogs not able to travel – made 2 m. and camped. Sent Alec on to break track for tomorrow. Valley very good and grade easy.

23rd Saturday – Fine morning – went to end of broken track, and managed 14 m. farther – camped at noon 26 ½ m. from the Cache. In p.m. went 3 ½ miles ahead with Alec, and found fine valley S50E for about 25 miles ahead. Through ice twice today.

24th Sunday – Left stationery box and sextant & rifle at camp No. 10 and returned 14 miles. Very heavy going, the track almost completely snowed up.

25th – Some snow fell last night but this a.m. fine & warm – too warm for travelling – returned to cache and camped there – track completely obliterated and ice considerably overflowed. Went through once.

26th Tuesday – At cache. Alec & Johnny broke track 7 miles down river. Hanington & Petit Jean returned in evening with one load of salmon, having left the other at mouth of river. They had very heavy snow and bad going, having to "double up" all the way.

27th Wednesday – Sent Alec & Johnny back with 8 dogs for the remaining load of salmon – heavy snow falling all day.

28th Thursday – Fine day. Alec returned in p.m., accompanied by 3 H.B. Co. trains, 8 days from Fort George, bringing 800 salmon and a few ictas [Chinook jargon for "things"].

29th Friday – At cache overhauling loads & reloading for a start tomorrow.

30th Saturday – Left the cache at 6.50 a.m. with 2500 pounds of provisions etc. on six trains. We have about enough salmon and grub for one month. Went 16 miles up river, and camped at the Forks. Surveyed a mile up the South fork, the general bearing of its valley is S45E or parallel to that seen on the 23rd, and separated from it by a range of bare rocky hills about 3000 ft. high [see Figures 19, 20, 21 and 22].

31st Sunday – Went 10 ½ miles up river (to camp 10) found the going very heavy, our old track being completely snowed up and much water on the ice. Camped at 3.20 p.m.

1st February Monday – Went 8 ½ miles up river and camped, No. 11, at 3.20 p.m. The snow is too deep for our heavily loaded trains to make any progress through it. Weather has taken a decided turn for the better. Sharp frost at night.

2nd Tuesday – Left party in Camp No. 11, and started ahead with two men and a week's grub (on a light train) to break track. Reached the falls at 2.30 p.m. (44 ½ miles from cache) and climbed an infernal hill on S. side, 600 ft. high and steep as a house – got a good view up valley (about S45E for 20 or 25 m.) but found it impossible to make a portage on that side. Went a mile back down the river and camped at 4 p.m. Camp No. 12. Snow in woods 4 ft. to 4'6" deep but just enough crust to make good snowshoeing. Both my feet very sore from the straps, and a slight touch of 'mal de raquette' in right leg.

3rd Wednesday – Climbed hill on N. side of falls, but found no chance to make any passable portage there, so went round the face of the hill & found it possible to get along there. Returned to camp for dinner, and found the balance of party just arrd. Took all hands along in p.m., and by bridging and grading made a good portage 65 chains long. Broke track to the upper end of the canyon, a mile more, and returned to Camp 12 at 5 p.m.

4th Thursday – I started ahead again (leaving party to make portage in two trips) and found another canyon which required a portage of 20 chs. Made 10 m. altogether and camped (No. 13) at 3.35 p.m. The snow on ice is fully 2 ½ ft. deep, and very heavy going. Heavy snow all day.

5th Friday – This a.m. came to another infernal canyon which fortunately was passed with only a little bridging. The snow getting deeper and deeper and falling fast still. The dogs pretty well played out. Made 10 m. to Camp No. 14.

6th Saturday – Snowing all night and this morning till dinner. Very warm, and awful travelling, the valley is narrowing up fast (being here only ½ m.

wide) and the stream is becoming smaller. After dinner finding the dogs could not travel I left them, and broke track a couple of miles ahead (8 ⅓ m. in all today) and then came back to the forks of the valleys to camp. Fired at a moose with revolver but the ball failed to connect. The valley I have been following continues to the S.E., but the main stream comes through a gap in the high bare rocky mtns to the East. Very warm and overcast at night.

7th Sunday – Took train as far as we broke track last night, and walked ahead 3 ½ miles – passing a 200 ft. fall on the way. Here the stream forks, and I followed smaller fork (of the stream) 6 miles East, where it ended in a semi-circular basin of high bare mountains and glaciers. Returned to sled, and found rest of party camped there. Very warm all day and almost impossible to get the dogs along, even on beaten road.

8th Monday – Overcast and very warm all day, with frequent falls of light soft snow. Hanington went 5 m. from the forks I was at yesterday following the larger stream; but it became more rapid and much smaller, closed in on every side by high glacier-covered mountains, and offering but small chance of getting through that way. The great altitude reached (3600) also deterred me from trying any farther in this direction, as it is evident the good pass spoken of does not exist here. I therefore decide to return 50 miles to the forks passed on the 30th ult., and to follow that valley. I plotted the survey as far as we have gone, and make the Lat. of this camp (no. 15) by D.R. 54° 08' North. Have had no opportunity for some days to make any solar observations.

9th Tuesday – Very warm last night and continuing till noon. Returned 18 miles to Camp 13, and camped there at 3.10 p.m. Weather cool and clear (for a wonder) at night, and new moon visible.

10th Wednesday – Went 17 m. down river, and camped at Camp No. 11 at 5 p.m. Very warm day, and a great deal of the track snowed up, and covered with overflowings.

11th Thursday – Went 11 miles down river, and camped at 3.30 p.m. – very heavy going, and much water on ice.

12th Friday – Went down river 7 m. to the Forks, where camped at noon. After dinner overhauled cargoes and counted the salmon. There are 604 left, thus being about 70 short, which have been stolen by dogs and men. Wrote to Mr. Fleming, and enclosed to Bovill, asking him to forward it to Quesnelle.

13th Saturday – Sent Petit Jean and Tsaiao back to Fort George with Hamilton's and Seymour's trains. Sent Alec with Johnny and Ahkoh (hauling a small sled) up the river to break track. Fine and very warm day.

14th Sunday – Started this morning with three trains of dogs driven by Hanington, Hassiack and myself, with about a month's provisions for men and dogs. Found Alec's camp 3 m. up river, he being away making portage to avoid the falls. In p.m. he returned, and hands graded the road to the top of the hill. Very warm day, but track good – the portage should have been made farther East than where we went, much of the hill being thereby avoided.

15th Monday – Alec started again ahead with the same men, and we took half our loads across the portage 4 miles and returned to Camp 16 in the afternoon. Fine day, but very mild.

16th Tuesday – Took the balance of the load across the portage, and loading up at the S. end, went 7 miles up the river to Camp 17. The track is tolerably good, in fact we could not move along at all without it, but owing to the entire absence of frost at night it is not hard enough to bear the dogs up properly. Mild day, some snow, rain at night.

17th Wednesday – Another warm day, with rain in p.m. Went 12 m. up river to Camp 18, stopping early (as Alec had done) on a/c of wet – the snow is almost slush and the dogs can scarcely keep the heavy trains in motion. My snowshoes completely "gone in" in spite of repairs last night.

18th Thursday – Another warm morning, everything wet through and miserable. Went 11 miles up the river, which still has a good valley and keeps about the same course. In p.m. came to Alec's camp – he has run out of grub, and made 8 m. of track up the river while waiting for us. The dogs have all they can do, with our help (pulling and pushing the sleds) to travel at all, and we frequently have to leave them, and go over the road again, then going back for the trains. I never saw such miserable winter travelling.

19th Friday – Went to end of track 8 m. up the river, and camped at noon. After dinner, started Alec and the two men ahead, I going 5 m. with them. The valley forks about 3 m. from camp, the river coming out of a gap in the main range N50E, and a small creek coming in from S20E. The pass is very distinctly marked, the high bare rocky hills coming down bluff to the valley on each side, & the valley looks low, thus promising well [see Figure 23]. It is to be hoped we are on the right track at last. Clear bright sunset and moonlight, 'hoffentlich' [I hope] a fine day and cold tomorrow.

20th Saturday – Heavy snow falling most of the night. Did not leave camp until 9 a.m., but managed to make Alec's camp of last night – 7 miles – by noon. While at dinner, the snow turned to rain, continuing all p.m. and evening. Had my snowshoes laced over, and spent the afternoon trying to keep our blankets &c dry. What a climate!

21st Feby. Sunday – Went 11 miles up the river, entering the mountains at the 9th mile. Heavy snowstorm all the morning, but cleared up a little after dinner, not sufficiently, however, to enable me to see any part of the valley or surrounding hills – Very bad going for the dogs.

22nd Monday – Met Alec at 9 a.m. and went 7 ½ m. up the river, making several portages to avoid open water. The current here is swift, and the rise considerable. There are several old winter camps on the river, probably made by Indians. Decided to keep all the party together now, as the track is liable to be snowed up, and thus rendered useless through the night, and the pain (in distance) is not worth the extra work. Weather still mild, tho' improving somewhat on the last few weeks. While camp was being made, went ahead a couple of miles with Hanington to break track for tomorrow morning.

23rd Tuesday – A bright clear morning and tolerably cool. The sun rising showed the valley ahead of me distinctly, and gave promise of a low and easy pass – a hope, however, too soon to be dashed to the ground. The stream commenced rising fast with a 50 ft. fall at one place (bringing the altitude to over 4000 ft.) when camped at 9 miles. Went on with Hanington, and found a forks on the stream ½ m. above camp – the larger coming from S.W. and the other from N.E., both through canyon and rising precipitously. We climbed a mountain right before us alt. 5400 ft. and about 2000 ft. more to the top but the p.m. was so misty nothing could be seen, except that I imagined the country low to the east....this proves the pass entirely impracticable for Ry. purposes, but as it is the only possible chance of passing the mountains I determine to proceed and endeavour to gain some topographical knowledge of the other side. On way back to camp, made track wh. will be available for portage tomorrow.

24th Wednesday – A cold a.m. at last! H. and I continued the portage to the top of the hill (alt. 5400) about 2 m. from Camp 24; and all the load was brought up in two trips, when camp was made in the same place. After dinner, went ahead with Hanington to explore, finding a lake at ½ m. Followed up this to the east, and after passing through 3 others (all in the same direction) were well pleased to find one whose waters empty the other way. Followed down the creek a mile, to satisfy ourselves we were *really* over the divide. Snow and mist prevented any view, but the question of the *passability* of the Rockies by this opening having been decided in the affirmative, we returned to camp much elated [see Figure 24].

25th Thursday – Left camp at dawn, everyone being anxious to see 'the other side'. About 6 ins. of light snow fell last night completely filling up our track, but under the circumstances that was thought nothing of. Crossed the summit of the Rocky Mountains at 9 a.m. and took a parting look at British Columbia – long may she wave! And may it be longer still before I see her inhospitable rock-bound coasts again. Went 12 ½ miles delighted with the 'down grade' – The country is still high and broken, with no regular outline; the valley of the creek we are following being ill-defined and cut up by lateral vallies & rocky ridges running in every direction. The whole country has been burnt over within a few years, and presents a most desolate appearance. Saw recent Indian cutting this p.m., the first on "this side".

26th Friday – Light snow fell in the night, and the warmth of the morning made us oversleep ourselves; did not get up until 6 a.m., instead of 4 o'clock, our usual hour. A couple of miles from camp came to a canyon, the entrance to the upper end of which was completely barred by a perp fall of 200 ft. in height – had to go back a few chains, and make a portage over the hill on the N. side nearly ¾ m. long, with a very bad and steep descent to the river. This occupied so much time, that we only made 5 ½ m. by 4.00 p.m., when we camped. The snow on the river is very deep and soft, being apparently freshly fallen, but in the woods there is a visible diminution in depth – it being only about 18" where we camped. Went ahead with H. and made a mile of track, the valley still continues East, will however shortly turn to N. or S., as it is closed by a high range bearing across it. The bare mountains still follow us on each side, but at some distance from the river – they are becoming much lower and more wooded. By an obs. of the sun at noon yesterday, I made the Lat 54° 07' N, or about 55' N. of Jasper House, wh. probably bears about S20E from us.

27th Saturday – The going began to improve this morning, the ice having been sunk by the weight of the snow, and the overflowings frozen up again. Made 15 miles by camping time – after dinner passed thro' a small canyon with a 20 ft fall at the upper end, and shortly afterwards came to a branch of the river coming in from the South and West. I went about a mile up this – it runs thro' a small canyon, with perp walls of rock 150 ft on each side, and several small falls – the body of water is about the same as in the fork we are following. The united stream does not increase much in size, but makes a big bend to the N, and seems to be closed in by a high snow-capped hill about 5 miles from camp. Unless it makes a decided turn to the East soon, we will have to leave it, and strike overland for the Athabasca.

28th Sunday – Got off early this morning, and having good going, made 8 m. before dinner – the river here turns to N.E., and shortly afterwards opened out to the East, the current is more rapid, and much open water gave us trouble as portages had to be made to avoid it. Very warm in afternoon and snow rather sticky – I had a bad touch of "mal de raquette" in my right foot which is very painful. Made 15 m. in all, and camped with a good prospect of valley to the East – cold at night and bright.

1st March Monday – Off at 6.15 this a.m. – our earliest yet, the days are much longer now, and the morning the best time for travelling as the sun has great power after noon. The general bearing of the valley about N70E all day – came to a prairie a couple of miles long with several Indian camps on it and marks of recent cutting – camped at 4 p.m. having made 13 ½ miles. All the snowshoes in a bad state & require fixing. Very warm in p.m. and heavy going – the valley closes in ahead of us with high timbered ridges crossing about a mile below camp – river appears to turn to the S.E. – very rapid current and a good deal of descent, but the baroms all wrong on a/c of high wind.

2nd March Tuesday – High wind from S.E. with snow squalls – in fact, a regular March morning. Cleared up towards noon, but remained cold. Very bad going, the snow deep, and the ice much overflowed, only made 11 ½ miles. The river is falling rapidly, and the general bearing of the valley (if that can be called a valley, wh. is little more than a tortuous cleft in the sandstone rock) is abt, S60E or S70E – this would lead to the conclusion that it is the R. de Baptiste we are following, and which will lead us to the Athabasca about in latitude 54° 04'. Have had no oppy. for an obsn. lately, but make our latitude (by D.R.) about 54° 14' 30". Very cold at night, and barom. rising ever since we camped.

3rd Wednesday – Went ten miles down-river – very heavy travelling, and all the dogs pretty well "played out". At camp, several old Indian lodges and a horse trail distinctly visible. The river here makes a big turn to the North, and I begin to suspect we are on the Smoky R. and not the R. de Baptiste – if so, this is probably where the Indians from Jasper House strike the river, travelling with pack animals. Will follow it, however, a day or so longer, as the direction may prove more favourable. Cold day, but pleasant.

4th Thursday – About 3 m. from camp saw a low valley to the south – and sent Alec up to explore it, it turned out to be bad going, and too much to the S.W. to be useful to us. A few miles further down found an old snowshoe track going down the river, evidently an Indian hunting, moose tracks also plenty. After camping, went a mile further with Hanington, and finding the valley of the river makes a turn to the N.E. (or farther still away from the Athabasca) I decided to leave it at once, and cross the country to the latter river. Cloudy night, but no snow.

5th Friday – Remained in camp all day, to give men and dogs a rest which we all badly need. As the dogs will not be able to haul the sleds over the rough country I expect to find between here and the Athabasca, I made a cache leaving everything we can possibly do without [see list at end of book] and prepared to start packing, having 16 days grub for all hands, and 110 salmon for 11 dogs. Plotted up the survey, and calculated the distance across on a S.E. course to the Athabasca to be about 40 miles. Made an obsn of sun on meridian, and determined Lat. At 54° 23' North (this agrees with the D.R. by plotting).

6th Saturday – Put some blankets and the salmon on the dogs, and carried the rest of the things ourselves – took Alec's sled with a few things on it, to try if possible to get it through to the Athabasca. Followed down the Smoky R. a mile, and turned off to the S. up a 30 ft. creek; after another mile, found this going too far south, and struck off to the S.E. over the hills. All the country is very broken, high ridges parallel to the river, and deep valleys intervening, covered with a dense thicket of small black pine, through which it is almost impossible to force one's way on foot, to say nothing of taking a dog train. Managed however to make 7 ¼ m. by camp time. Half a mile before camping we crossed a high ridge running to the N.E. (alt, 4500) and saw across another deep valley an equally high hill, right in our course. Trees laden with loose snow, and everything wet through, the sun being very powerful in the afternoon. H. has a bad touch of "mal de raquette" – my foot somewhat better.

7th Sunday – Descended to the valley, and at a couple of miles from camp came to a 300 ft river running to the N.E. Followed it 2 m. south, and turned off to the S.E. up a 30 ft. creek. At the mouth of this creek found a saw-log on a drift pile, and accepted this as a pretty conclusive proof that this is the much-wished-for Athabasca. The river is not, however, so large as I expected to find it, nor does it agree with the D.R. as to longitude – Very warm afternoon, & heavy snowshoeing.

8th Monday – Followed up the same creek 7 ½ m. in a southerly direction; it is very crooked, but still much better travelling than through the thickets on the banks. Climbed a hill after camping, and saw the valley we are in turning away to S.W. a couple of miles further on. Mild day, but the going somewhat better. Saw two moose, but of course the rifle was not at hand.

9th Tuesday – Left the creek at the bend seen last night and climbed a high ridge (5000) from top of wh. saw spurs of the Rockies bearing S.W. The rest of the day climbed over successive ridges, all running to N.E. and S.W., and divided by small streams running N.E. At camp (barom, 5100) a 10 ft creek runs ½ m. east, then turns due north; some snow in p.m., but pretty good travelling in the woods, as they are more open than I expected.

10th Wednesday – Followed down the creek we camped on, for ½ mile, and then up a small gully to the S.E. Saw several blazes where left creek, and signs of a horse trail, running up the creek. Climbed the ridge to altitude 5400, and saw before us a wide valley with a higher ridge on the opposite side, above which the mountains near Jasper House (suppose R. a Miette and Roche a Perdrix) were clearly seen 30 or 40 m. to the south. Descended to the valley, wh. is open, partly prairie and partly muskeg, and camped halfway across it. Got a good view of the main range of the Rockies, running N70W, with the same valley we are in now continuing along their base and separating them from the high ridges we have crossed. Bright and warm day, very high west wind in the open.

11th Thursday – A short distance from camp struck a 20 ft creek running down the valley seen yesterday to the N.W. (this is probably the headwaters of the other branch of the Smoky) & followed it to its head. Here I found the same trail I saw yesterday morning, and followed it across the divide to a creek running S.E. I take this to be the valley followed by the Jasper House Indians going to Smoky River. The trail is well cut out & blazed. Saw the same mountains to the S.E., about 30 or 35 m. distant. Warm p.m., but going good.

12th Friday – Some snow fell last night, and continued at intervals during the day, with high E. wind. Followed down the creek (which rapidly developed into a 200 ft river) in a S.E, direction, but the snow very deep on the river, and heavy travelling. A very disagreeable, regular March day. Saw occasional signs of the trail along the banks, & several old Indian camps.

13th Saturday – Finding the river turned away to the N.E. (as they all appear to do, in parallel valleys) I left it a couple of miles below camp, and resuming the old S.E. course, struck across the hills – passed over some high ridges, and an undulating country with open black pine, and a few horrible thickets for 10 ½ miles, when camped.

14th Sunday – An early start this a.m., and a pleasant day for travelling. After dinner saw signs of an Indian trail, and shortly afterwards came to a 100 ft river, running N., and which we followed up for 1 ½ miles. Here it turns away to the S.W., and the valley could be traced away to the W, where it heads in the mountains. Both yesterday and today very hazy, so we could not see much of the country we are travelling in, nor of the mountains we are steering for.

15th Monday – A high brule ridge to the front turned us to the south this morning, and at a couple of miles came to a lake, on which the going was very good. Another mile and a half farther, being on a brule hillside, we were delighted to get a good look at the Roche a Miette (opposite Jasper House) bearing S10W and apparently 20 or 25 m. distant. This is a good landmark, as the peculiar shape of the rock (sketch) standing apart from the other mountains renders it unmistakeable, we are so far on the right course, and everybody is encouraged.

Passing through 4 more small lakes to the south, I would have proceeded direct for Lac a Brule (wh. must be between us and the R. a Miette) but the rocky and broken look of the country on that course deterred me; I accordingly turned on the old S.E. course (which has never failed us yet) and climbed over a high ridge, partly burnt and part green, from the E. side of which the Athabasca valley appears just at our feet. Camped about 4 m. from the valley – Very heavy travelling in the open burnt woods.

16th Tuesday – About 4 m. from camp found a well-beaten snowshoe track running along the benches above the Athabasca, which is distinctly and unmistakeably close at hand. Three miles farther came to the valley of Freeman's River, wh. we descended to Lac Brule just where the big river runs out. A cheer from all hands announced that the toils and perils of our journey were nearly at an end, and, taking off our snowshoes we made good time to the Fiddle River Depot, built by Mr. Moberly. But our hopes were here dashed to the ground; for we learned from the Indians camped there (and some of whom have taken possession of the one remaining house) that the Company's post at Jasper's is abandoned, and that they themselves were unable to supply us with the provisions we are so much in need of. In the evening, had a 'talk' with the noble red men; which ended in their producing some dried meat, but in small quantities only. Got a good meal of rabbits for supper and some scraps for the half starved dogs. Turned in early – "carpe diem".

17th Wednesday – Decided to rest at the depot today, and as we cannot get any grub for the dogs, I arranged to leave them in charge of the squaw – who promised to take care of them till they could be turned over to us (or to the company) in the spring. After a little more talk, and the sight of my gold (*auri sacra fames*) they produced some more dried meat – and learning that one of the men had half a mountain sheep at Jasper Ho. [see Figure 25], I at once arranged with him to go up for it – we got two horses, and rode up together – there is nothing left at the Post but a little powder and shot, & 2 tin cups – the latter I seized, as we are badly in need of them. On overhauling our provisions on my return, I find there is enough (with strict economy) to take us to Lake St. Anne's, about 200 miles distant – so made up the packs, and ordered a start in the morning.

18th Thursday – Took an early leave of our worthy host, who presented each with a pair of fancy moccasins, and started down the lake. Very high south wind rendered it almost impossible to keep our feet on the glare ice, so we had to skirt the shore on snowshoes. Went down the river, where the going was good, for 9 ¼ m. below the lake – making track survey as usual. The immediate bank of the river presents unusual facilities for a good line.

19th Friday – Got off at 6 a.m. and followed the river 14 miles more. Here we turned off to the S.E. and, climbing a high bench, struck the H.B. trail at ½ a mile. Followed it a short distance, and camped.

20th Saturday – Very heavy going on the trail, and difficult to follow it; only made 8 ½ miles, and pretty well "played out". We camped just beyond a large lake, wh. lies about a mile north of the trail, and the N. shore of which looks very favourable for a line. I would suggest following the Athabasca to about where we did (or a little lower down) and then striking across, 4 or 5 m. north of the H.B. trail toward the Macleod. Warm in p.m.

21st Sunday – Struck Moberly's trail early this a.m., and found it much easier to follow than the old one we left. Followed down along the Mcleod; crossing the surveyed line in several places. Very warm in p.m., but a sharp snowsquall at camping time was succeeded by intense cold. Made 12 m., deep snow.

22nd Monday – Got off pretty early this morning – 5.05 a.m. but only made

14 miles, heavy snowshoeing being the cause. We are, however, nearly across the McLeod Portage – took last look at the Rockies for some time, I hope – also got a good view to North, confirming my ideas of a more northerly line than the present one. Very cold all day, and at night.

23rd Tuesday – About 10 o'clock this a.m. met the H.B. Co. outfit going up to Jasper House with three loaded sleds – they are eleven days out from St. Ann's, had heavy snow and bad overflowings – I managed to get a few pounds of pemmican & a little tea from them & we had a square meal in consequence. Their track is of great assistance to us, and we made a good afternoon's journey.

24th Wednesday – A beautiful day, & the track in good order – we went along gaily, but were surprised to find the H.B. people had not come by the ordinary trail from Dirt Lake – followed their track, however, down the river, as they probably have made a short cut somewhere.

25th Thursday – Another day down the river, 21 miles, still following the track, and wondering why it doesn't turn off – it is now thought they may have made a short portage direct from Dirt Lake; and as the river still runs tolerably well to the East (much more so than shown on our map) and are not losing by following it – very warm in p.m.

26th Friday – As the river began to turn away to the N., and finally the N.W., I got disgusted at following it so far round, and about 9 o'clock decided to strike direct for Dirt Lake (about 25 m. distant). Found the travelling very bad in the woods, but managed to get 9 m. from the river by camping time.

27th Saturday – Made 11 miles today, some good and some horrible going, struck a creek in p.m., running Easterly, which I take to be Dirt Creek, but as it is so small I fear we must be some distance above the Lake. Made a careful division of the provisions this evening, & find ourselves on very small allowances, were fortunate enough to procure a couple of rabbits wh. helped to fill our half-empty stomachs.

28th Sunday – About 5 miles from camp this morning came in sight of a large lake, and going down to it, half a mile further on, decided it to be Dirt Lake – having thus made a very lucky strike for it, and just about the distance from the river I estimated. After dinner I crossed to the north side, & found the H.B. people's track along the ice – they have therefore gone this way and made their portage lower on the Mcleod than I did – well pleased to see "signs" once more, we pushed on, over the 8 m. of lake (about half its length) and camped 3 m. below it.

29th Monday – Following down the creek and making several portages across points, we reached the Pembina shortly after dinner; three miles down this, came to the horse-trail crossing, and went on (leaving the river) to the first small lake – here we lost the track (which is much snowed up) and going out on the lake were delighted to see a smoke in the woods half-way down. Although it was sun-down, we pushed on towards it, and found a Stony Indian camped there – from him I procured a dozen rabbits, otherwise we would have gone supperless to bed. Very warm in p.m. & a regular spring day.

30th Tuesday – Off early, and soon found the Indian and his boy following us – Bargained with him to break track for us, and to carry a letter to the H.B. post to notify our arrival, as we would probably come late. Very warm in p.m. and snow very sticky; we are <u>almost</u> played out with hard travelling & want of grub – arrived at the Co.'s post, Lake St. Anne's about 7 p.m. and recd a welcome, and a good feed of whitefish & potatoes from Mr. McGillivray, the officer in charge. Very glad indeed, to be at the end of the hardest part of our journey – having had a most wearisome tramp of 217 miles from the Fiddle River Depot. Gave the men a good blow-out of all the good things we could raise – and turned in under a roof once more.

31st Wednesday – Resting at Lake St. Anne's, and eating as much as we can. Alec and H. went to the settlement, and procured some eggs, butter & milk, all great luxuries to us in our impoverished condition. Mr. Kirkness (who is to succeed Mr. McGillivray at this post) arrd from Edmonton in p.m.

and we arranged to return at once with his teams – got some provisions from the store & some ictas [things] for the men.

1st April Thursday – Left Lake St. Anne's this a.m. with one cutter and 1 horse-sled – and went about 32 miles towards Edmonton, when camped – very warm day & heavy rain at night.

2nd Friday – Disagreeable stormy morning – a good deal of fresh snow & drift. Reached the Fort at Edmonton at 10.30 a.m. and got a room from Mr. Hardisty to stay in. Procured an outfit of clothes (wh. we sadly need) – called on Col. Jarvis in p.m.[2] H. and I get our meals at Mr. Hardisty's house – Alec at our room. The men drawing rations & living in Company's house. Several bad attacks of cramp in my leg since last night, can scarcely walk at times. Storm ceased at night.

3rd Saturday – Took a good rest in bed this a.m. and had our breakfast in our own room. Drove out to Norris (at Long Lake) with the Colonel – and tried to arrange with him to go with us as far as Carlton. He came in to the Fort in the evening and I arranged with him to accompany us as guide etc. for $250. Made application to Mr. Hardisty for a dozen horses, and some flat-sleds & carrioles. Cramps still very bad – & Hanington suffering from diarrhoea.

[There are no diary entries after April 3. The rest of the journal consists of financial accounts and sketches. The next extant diary commences on May 19.]

May 19th – About 8 miles from camp came to the 1st crossing White Mud River, and took dinner a couple of miles beyond 2nd Xing. Had to make a long detour to Tatogan to avoid bad roads. Camped at Shannon's, and after supper H. and I drove on to the Portage, hoping the stage may leave tomorrow for Winnipeg.

May 20th – Reached the 1st crossing abt 8 miles from camp and took dinner a couple of miles beyond the 2nd xing White Mud River – had to make a long detour to Tatogan to avoid bad roads. Camped at [...] [This entry repeats the previous day's and ends at this point, and a diagonal line is drawn through the entry.]

May 21st – Left the Portage at 7 this morning by Tait's stage. Arrived in Winnipeg 7 p.m., met several old friends but found the Macaulays and St. John's left last Tuesday for Canada. Stopping at Crown's 'Grand Central' hotel. Fine, warm day.

Saturday May 22nd – Reported to Ottawa and recd instruction to prepare plans and report – Fine warm day, lunched at the barracks, and called on McDonald's afterwards. Got trunks to hotel, also a great many letters. Telegraph to wait.

Sunday 23rd – At Fort Garry – In office all day – spent evening at Mrs. McDonald's. Norris, Alec & Johnny arrived with the horses & carts, and camped a couple of miles out of town.

Monday 24th – Finished report and plan, of which Hanington will make a tracing to send tomorrow to Ottawa. As Paymaster Nixon is not here I drew on Bain and made some payments on a/c to men.

Packed up in evening, and told Crown to give H. my room while I am away. H. at dance at barracks.

Tuesday 25th – Left Ft. Garry by stage at 4 a.m.

CHAPTER 4

Excerpts from C.F. Hanington's Reminiscences, 1928-1929, Describing the 1875 Expedition

Hanington described his professional career in detail in a series of letters to John Hosie, British Columbia provincial archivist, written between June 1928 and April 1929.[1] He titled them his "Reminiscences," and they convey his dry sense of humour (and are chaotically typed). This excerpt covers the period of the Smoky River expedition, including the section from Quesnel to Prince George that is missing from Jarvis's diaries.

Next day was spent in sending off our horses, paying off the packers, and getting ready for the trip to Quesnelle by stage. Everything was lovely in the way of being forwarded by the B.X. as usual, and as the 100 Mile House was just like all the other stopping places, with plenty of water for washing, plenty of Bass Ale for drinking, Bunch Grass beef steak for food, with vegetables, and no end of pie of various sorts, we had nothing to do but enjoy ourselves in a very quiet way. It was a very pleasant drive behind the well-kept and well-bred horses of the stage line, the B.X. Express, and I do not recall anything out of the usual in the appearance of Soda Creek and the surrounding country. It was Quesnelle that dwelt in mind, the jumping off place from which we started on an unknown trip through a decidedly unknown country.

We reached Quesnelle, put up at the hotel of Ben Gillis, and found that he originally came from P.E. Island. As Jarvis was born in the same part of Canada, and as I came from New Brunswick, a few hours away, with the straits of Northumberland between us, we concluded that we were all three inhabitants of the same country, and immediately began to be chummy. We had expected to find the Fraser River frozen, or nearly so, and intended to travel over the ice to Fort George and onward from that H.B. Co post, as soon as we got dog trains and supplies with a winter outfit and were ready to start. In the meanwhile we were comfortable, and while waiting put ourselves into training by walking every afternoon, hard and fast up the road to Caribou, beginning at two miles out and back, and getting up to five miles each way, and ending by running the ten miles and enjoying it. The mornings were spent in hunting up dogs and outfit. Chum was the first dog we got, and I believe he was really the worst we ever got. A dirty yellow, he was, and his character was at least as dirty as his hide. The finest, most Christian dog in the lot was Marquis, who was a sort of half Huskey, strong, willing and well trained. Alec McDonald was employed as O.C. Dogs, and he certainly had his hands full with Chum, although he knew how to handle that sort of dog. We could get no pemmican, of course, but got beef by the quarter and cut slices from it, and filled two sacks with the meat, leaving the bones. Mr James Reid was the postmaster, express agent, and kept a supply store as well. He lived in his own place not far from the Gillis Hotel, and did all he could to help us get ready for the trip. A bunch of prospectors were coming in from the Omineca country, and they were a novelty to me, as most things in a mining district were. The bar was in a very large room, into which the front door opened, and a large stove stood in about the centre of the large room. There were tables and chairs, occupied in the evenings by these returned prospectors. And I found I had much amusement and interesting stuff in their conversation, into which I, being a Chechacko, did not join. Soon after the old prospectors left by B.X. stages for the coast. Mr Justice John Hamilton Gray arrived from Barkerville, where he had been holding court. He was the father of Jack Gray, who was a member of Division M, as you know, and Judge Gray had been sent to B.C. from Ottawa, as judge of the Supreme Court of B.C., and of course we were delighted to see him, as he and his family had been very hospitable to us in the spring, and in addition to that, they were also from New Brunswick, he having represented St. John N.B. in the House of Commons at Ottawa. The Judge remained at Quesnelle for two or three nights, and that added to our enjoyment, as he was a most interesting and educated man, with a great and useful knowledge of Canadian matters, and of the world in general.

After Judge Gray left for Victoria, the days became more monotonous. We kept up our heavy exercise, but worried because the river remained open, and time was going at about the same speed as the open water. We were now ready to start, but no ice appeared upon which we could safely travel. At last the chief decided to give up the idea of travelling on river ice, and go to Fort George over land, by the telegraph trail as far as the Blackwater River, and thence via a trail made by H.P. Bell in the summer before. None of us had seen either of these trails, but we had courage and hope, also enough grub to take us to Fort George at any rate. The Citizens of Quesnelle turned out in a body to give us a send off, and their remarks were not very cheering or hopeful. "Goodbye old chap. We will never see you again but your frozen bodies will turn up after the snow goes off." This was a very common sort of farewell, and we were glad to get across the river (by scow) and out of sight of the waving, shouting, well meaning friends. There was enough snow to make good travelling on the flat portions of the country, but Oh, what a trail when we struck side hills and got off the flats. The Telegraph trail was built by a Telegraph Co, who proposed to put up a telegraph line to the Far North, and carry it across Behring Straits to Russia.

We did not get away from Quesnellemouth until December 9th, 1874, and although it looked like a long way to Fort Garry, still we really thought we might get there. At least I hoped so, and imagined that Jarvis also had at least as much hope bottled up. Oh, I have omitted to tell you about the Telegraph Trail, over which a telegraph line was to be built by (of course) an English Company, who had really put up many miles of wire, had opened many offices about 20 to 25 miles apart, and would ultimately have done as they intended, established a telegraph line of communication all over the world. Alas (there is generally an Alas in schemes that would work for the benefit of the whole world) the Atlantic Cable was laid, and proved to be a success, which astonished the scientific world, and broke the heart of the promoter of the telegraph line through the northern part of B.C. and across to Russia by the Behring Straits. Do you know, I still have much sorrow and sympathy for the man who thought out the scheme of connecting the East and West through, or by, such a service. However, I need not sympathise with the clever and hard-thinking man, because he has been dead so long that he will not appreciate my admiration and congratulations.

We crossed many small streams on bridges built by Indians, and made of poles lashed together by telegraph wire, which was to be found in almost unlimited quantities near at hand. It was rather novel to be crossing a stream over poles and wire swaying in the breeze, and looking anything but safe, but no accidents occurred. After we crossed the Blackwater River, over the same style of Indian (clever Indian) bridge, we struck the trail made by Mr H.P. Bell that same year. Of course, it was made for horses and mules, and was naturally kept above the soft and swampy places. Dear Reader, DID you ever try to drive a train of dogs along a trail on a side hill, and try to keep the toboggan from sliding down hill? If you did try and did not use all the oaths and then some you had never heard before, and failed at that, I, being polite, would only say that you are a liar. In B.C. there are THREE kinds of liars. No 1, a liar; No 2, a damned liar; No 3, a mining expert. Which are you? Answer in our next, please. Well anyhow, we were so fit and full of hope and energy, that we simply held up the sled to where it should be, sweated, and did the best we could, in an unexpected and entirely hard position, which could only be overcome by hard work, with slight traces of discomfort attached. We were TEN days getting to Fort George, and we had left every pound of stuff that could be spared along the trail behind us. I have never seen Bell's Trail since, that is why I am feeling so well and reminding the reader it is nearly 54 years since I saw the Bell Trail.

Well we got to Fort George on the evening of 19th December after one heluva ten days trip over trails that were never built for dog trains and apparently only existed in the imaginations of the so-called builders. No matter, they are all dead, and nearly forgotten by those who reaped the benefit from their work. At Fort George we found Mr Boville in charge, who was allowed in lieu of cash 5 dried salmon a week, ten pounds of sugar, and five whole pounds of meat per annum. Also in addition to his salary he was allowed to catch and hook all the game in the way of grouse and rabbits he could find. As he was an Englishman, and a gentleman, he was not immensely pleased with the prospect in store for him in the way of getting wealthy. From Fort George Alec and the Indian boy were sent back to pick up the stuff left along the trail by, and we three white men sat down and waited for dog trains, snowshoes etc to be sent down from the H.B. Co post further north. The Fort George was about half a mile from the Indian situation, which was further east. Outside of the H.B.C. post there were no buildings except Indian huts, but we were comfortably situated and well looked after, so far as Mr Boville was able to do it, and he certainly did his best.

Alec was a long time getting back, and only arrived late on the last day of December, having had no end of trouble on the trip. He got to Quesnelle on the 24th of December, got a canoe and an Indian, loaded the dogs and sleds into the canoe and got above the Cottonwood Canyon by water. Here, as the ice seemed strong, he sent back the Indian and canoe and tried to travel on ice. Alas it was covered with water and snow, and was impassable, as the slush stuck to the toboggans, and were too heavy to be hauled by anything but elephants. Alec doubled up the teams, took to land, got out of food, but found some Indians at the mouth of the Blackwater River, got dried salmon from them, hired another Indian to help, and got to Fort George, as I have stated, with the thermometer 49 below zero when he arrived, looking like a half-frozen, used up man.

The Indians from Stuart Lake had not yet arrived, so Jarvis decided that we should make a start, and get as far as a cache of provisions I had left at the foot of a waterfall, up the north fork of the Fraser, by canoe, in the autumn, after we had met Bell's party, and before striking across to the 100 Mile House via Green Lake. So we said goodbye to our genial host,

Mr Boville, loaded everything the sleds would hold, and started for the unknown, hoping to have the men, dogs, sleds and supplies overtake us at what was called Hanington's Cache. We had three white men, three Indians and three dog trains of four dogs each. It was very cold, 30 degrees below zero or more at nights, and as we had no tents it was not what you would call sultry, or even mild, at any time.

Fort George is situated at the mouth of the Nechaco River, which empties into the Fraser; there was one H.B. Post, and some distance away some Indian houses. Well as I think I have said, we were off at last, full of hope, but not feeling that two months supplies were going to last long enough to get us to our hoped-for destination. I speak for myself only, as I cannot guess what was in the mind of my chief, who said little but apparently thought a good deal. We were seven days getting to Hanington's Cache, to which I had taken some supplies by canoe in the autumn before. Here we unloaded and camped, but I was chosen to go to Bear River for dried salmon for the dogs. That was one heluva trip, the river having been flooded by the weight of the snow, every snowshoe track was made in slush, and at every step the slush had to be knocked off the snowshoe by a good-sized stick. To avoid this slushy stuff Quaw, the Indian guide and owner of the salmon, said he could make a short cut by cutting off the point between the Fraser and Bear River, like this. He had been across some 35 or 30 years ago, and knew the trail well, etc etc. We never did find any signs of a trail, but finally got to the cache, where we slept, and started back the next morning after loading the sleds with about a thousand pounds of salmon. We of course took the ice or rather slush on Bear River, and it was simply impossible to get along at all without turning the toboggans over every few minutes, and scraping the accumulated slush off their bottoms. It was that way until we got back to Hanington's Cache, and it really took more out of the poor dogs than a month's decent work would have done. We were away six days and had only done about ninety miles. I shall now add to what Jarvis wrote to Fleming (Chief Engineer) in his narrative of the trip.

I shall try and add some incidents that Jarvis did not mention, if I can recall any. It seems hardly fair to put Jarvis' narrative into my own reminiscences, but I still hope to get my own journal written in the winter of 1875 – 76 at Tête Jaune Cache, and that will cover everything I did and thought, I believe, some deeds and thoughts probably not worth recording. So far I have depended entirely on my memory of what happened 54 years ago, so perhaps I need not apologize for things forgotten. At any rate I want to assure the reader of this stuff that every word of the reminiscences is true, and although some incidents are hard to believe, yet still they are absolutely TRUE, and not exaggerated.

I see that my Chief mentions the fact that our dogs were frozen while hauling their loads and very hard work it was. That is what a tenderfoot would put down as an infernal lie, yet it is the truth without any exaggeration at all. The dogs were all provided with moccasins, as anyone with experience in winter travelling would tell you without shoes the dogs get snow jammed between their toes, and unable to waste any time by stopping, they naturally try to dislodge the agonizing bits of snow with their teeth. Naturally they cut their toes badly, and are soon unable to work at all. The dog shoe, or moccasin, is simply a small bag made of deer skin, with a running strand of deer skin around the top. The shoe is placed around the foot, and tied above the first joint, and it is a comfort to the poor dogs as long as it lasts, which is but a day or two, when it has to be replaced by a fresh outfit. However probably some veterans who have never been out of the civil service offices in Ottawa would be able to describe this all better, much better than I can attempt to do. At any rate when I unhitched my team one evening, I noticed that Marquis the Leader was walking in a stiff-legged manner, but I thought it was nothing but a streak of rheumatism. He was all right in the morning, but when we stopped that night Marquis was frozen and stiff, all the way up his legs. He ate his dried salmon, after it had been warmed in the red hot fire, and I noticed that he immediately got as near the fire as he could. In the morning his poor legs were done. He was unable to walk, so he was left behind, and the Chief lingered a little while and saw the end of a good dog, which came suddenly, although Jarvis said he hated to shoot the poor, honest, faithful worker. No, reader; I would not believe it myself, had I not had the experience,

Marquis was the first, but not the last, and I fancy they were glad to be out of their misery, rather than stay working for men who starved them (against their own wishes), beat them, trying to make the impossible possible, all of which the poor dogs did not understand. As Jarvis wrote, the coldest night was in the middle of January when it got down to 53 below zero. We had a self-registering spirit thermometer, and I kept the register. At noon, or shortly before, we stopped and got out the sextant and artificial horizon, which is quicksilver, kept in a lead container, and thus we got the latitude. If we wanted to check the time of day we had to take the sun's elevation at about 10.30 but that was only done if we found we were a bit out in the sun's elevation at noon. I am getting verbose, excuse me. What I started to say was that I picked up the quicksilver which was as I said in a lead case, and burned every finger on my right hand. The mercury was [remainder of sentence is missing from typescript.]

Jarvis decided after a while that we would never get through at the rate we were going, and an idea that if he and Alec went ahead with a light toboggan, carrying only a blanket and food for two, that the track they left behind would be frozen hard and that I could follow with the trains and balance of provisions etc. It was so, so we (the dogs and drivers and I) had a day's rest; next morning we were off bright and early, and caught up the trail breakers at noon, as we had a good, hard trail, and no slush to contend with. Alas, it snowed that night, and although a foot of snow gave us a warm night's rest, yet when we got up and started the fresh fall of snow made the slush conditions as bad as ever. Of course the food proposition was getting to be a puzzle, but at any rate the sleds were getting very much lighter, and so we went on as before, counting our paces, 40 paces to 100 feet, Jarvis doing it for half a day, and then Hanington doing it until evening. We had no tents of course, just a strip of factory cotton between us and the north wind, but we always had a large-sized fire in front of us, and as Jarvis and I slept under one blanket with another one over us we did not die, but were never warm for months. That episode mentioned by Jarvis, I mean the slide, was rather thrilling, in spite of the fact that thrills were not the sort wanted at that time. I remember that we washed at the summit camp, and I really believe that being a useless and uncalled for bit of work.

Some people seem to think that a bath a day is an absolute necessity. Only imagine, reader, we had not washed since leaving Fort George, and were not a bit dirty yet. How could we be? One does not perspire much in way below zero weather, and although I have heard of people getting warm at night by shivering, I have failed to get that experience.

I am cutting this short, but Jarvis' narrative will give you all the details I have omitted, I hope. We left everything we could at the summit camp. One dog, Captain, was left, and we packed all the remaining food, and the too few blankets on our backs, and struck away to the east by the south route, hoping to find something, or someone, before we saluted St. Peter at the Gate of the Great Beyond. I can only speak for myself when I say that I used to dream of roast turkey and plum pudding at the old home, and that I would have given one of my hands for as much bacon fat and bannocks as I could eat. I think now that I must have been rather hungry. We got to the Unknown Valley in the distance, which should be the Athabasca River, but if not, was sure to be the Valley of Death, so far as we were concerned. I wrote up my diary that evening, wasted no time upon dinner, as we had nothing to eat, and was almost sure that was the last of our wanderings. That is where I was wrong.

The narrative written by E.W. Jarvis C.E. covers almost everything, I think, but he does not mention the fact that after we had passed the summit of the Smoky Pass, I being at the rear, counting my paces and making notes in my notebook, noticed that Jarvis had halted, and was apparently thinking of something important. I hurried up to him, and said I hoped he was not thinking of turning back, although we had found the pass to be impracticable. That was exactly what he proposed to do; as he said, he was responsible for the lives of the party, and was in grave doubt as to whether we could pull through or not. I pointed out that he and I had no-one to take care of, that Alec was in the same position, and the Indians were not afraid, and that I personally would rather starve to death than turn back, and there were eight others in my family. So after a while Jarvis said "All right, we will risk it," so we went ahead.

We took the sun's altitude at noon every fine day, and got our latitude in that way, using a compass and counting our paces for dead reckoning. We carried a thick stick, with which we knocked the slush off our snowshoes at every step. This stick was carried, at the start, in the right hand, and the first pace was made with the left foot, one, two, and so on. At forty paces the finger of the left hand closed. That was one hundred feet of distance; again one, two etc until another 40 paces were counted, when the next finger closed, making 200 feet of distance, and so on. When the fingers and thumb of the left hand were closed that was 500 feet of distance, and the stick was shifted to the left hand, and the closing of the fingers and thumb of the right hand began. If and when that hand was done, the distance travelled would be 1000 feet, and a note was made in a note book. Of course after we left Smoky Pass and travelled through bushy country, which was at times very rough, one could seldom travel a thousand feet on one course, but had to set a new compass bearing, and make a note of the distance travelled on the last course. Savey [*sic*] vous?

When we reached Fort Garry, and after having a bath and change of clothing Jarvis and I walked around the hotel vèrandah. Jarvis said "How far do you make it?" and I replied after looking at my hands, which had acted without any orders, and closed the fingers at every forty paces automatically. Funny, eh? There were forty paces to the hundred feet, one hundred and thirty two to the mile, and as we had counted paces for eight hundred miles, deducting those made on returns from Head of North Branch and Fiddle River Depot, we two had counted over 10,000 paces during the winter [at the end of his letters to his brother he provides a total of 2,188,900 paces]. No wonder it had become automatic. At Fort Garry the Chief left me to plot a plan of the exploration, and went on to Ottawa. As Fort George, and also Quesnelle Mouth, Lake St. Ann and Edmonton were fixed points, I plotted the map at a scale of four miles to one inch, and as far as longitude was concerned it fitted the map of the same scale, exactly, but I had to swing the [typescript is unclear] further south, as we had made our latitude 54.20 and were some minutes astray, as it should have been about 54.08, which considering our means of getting the proper altitude of the sun at noon, with an artificial horizon of frozen quicksilver at times, and a rope stretched between two trees and placed in position with a hand level, was not so very bad, after all.

At any rate we did not starve quite to death, although it took several years to get over the result of overfeeding, after starving for so many weeks. Fort Garry was a small place in those days, and I could find nothing in the way of plan paper or tracing linen, on which I could plot the plan. However, finally I got a roll of wall paper, and used the plain side, which was rough in comparison to the proper plan paper, but served the purpose. I never saw E.W. Jarvis again for two or three years, as he was asked to return to B.C. and spend another winter in the Yellowhead Pass, locating a line down the Fraser to Fort George, and on to the west, and he simply pitched up the job and resigned. I shall finish this account at my next attempt, I hope.

At Fort Garry in June 1875 I was very busy, as my chief went on to Ottawa soon after we arrived at Fort Garry on the 21st June [May] as I had to get the plan done, before Jarvis got back. I was surprised one day to get a telegram from Mr. Fleming, Chief Engineer, informing me that I must report to G.A. Keefer at San Francisco, and that Jarvis was not returning. I also received a cheque, a bonus of 36 dollars, which the powers that be had figured in this way. Ten percent of salary for four months, as a bonus. As my salary was $90 a month at the time, I was presented with the munificent sum of thirty-six dollars. Jarvis, whose salary was $140 a month, got a cheque for $64, and our voluntary exploration had saved the government the cost of a survey, somewhere between $120,000 and $150,000, as there was a strong feeling at Headquarters that the Smoky River Pass was the very best and most feasible pass through the Rockies. It was a Liberal government in power in Ottawa at the time, and that is a specimen of their liberality.

I wonder what I have omitted that may be of interest. Oh, yes. An Indian named Johnny, who came from Stewart Lake, came along with us to Fort Garry, all the others except Alec remaining at Edmonton, to be sent back by the H.B. Co in spring time. The last I saw of Johnny was on a sidewalk with his feet in the gutter, one arm around the neck of a very pretty Cree girl, who looked as if she enjoyed the situation just as much as Johnny did. They did not understand one word of each other's language, but at any rate I said Farewell Kloshe Klahowya to Johnnie and his fiancée, and left. Johnny, of course, was being looked after by the H.B. Co, and looked fine. He came through to Tête Jaune Cache in the autumn of that year, and strange to say I was there, as I will tell you later on. Now I had to go to Frisco, via St. Paul, Chicago, Ogden, Sacramento to San Francisco, alone, and I simply hated the idea, BUT I WENT all the same.

P.S. I forgot to say that before we left Quesnelle Mr Jarvis was assured that at Fiddle River depot, which were winter quarters established by Walter Moberly, not far from Jasper House, we would find supplies of all sorts. Fiddle River depot is near the Athabasca, a branch stream, but alas, after we had left B.C. some official of the government had ordered those supplies sent to Edmonton, and there was nothing left for hungry men.

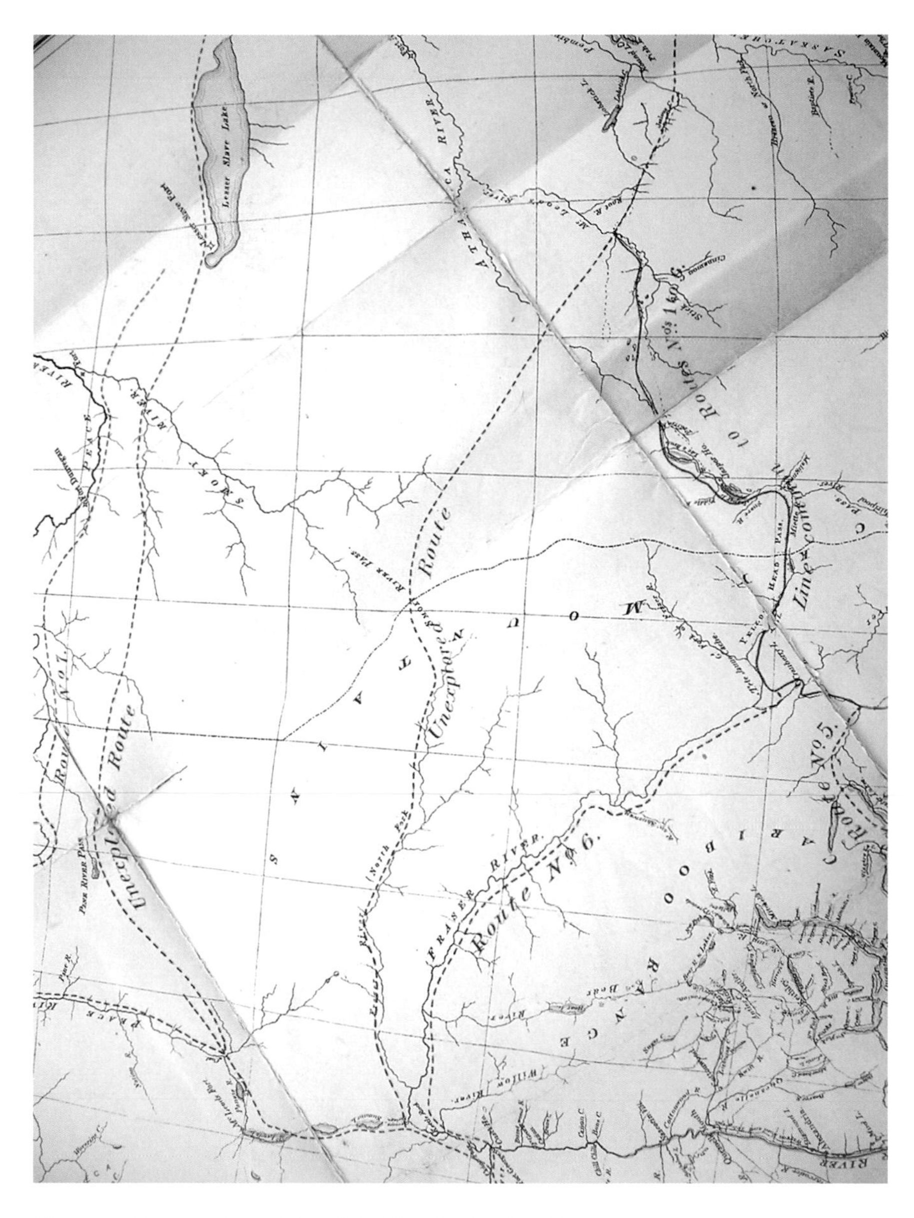

Figure 1. Excerpt from Map 8 in Fleming's 1874 Report. VANCOUVER PUBLIC LIBRARY, 385.09 C22FE

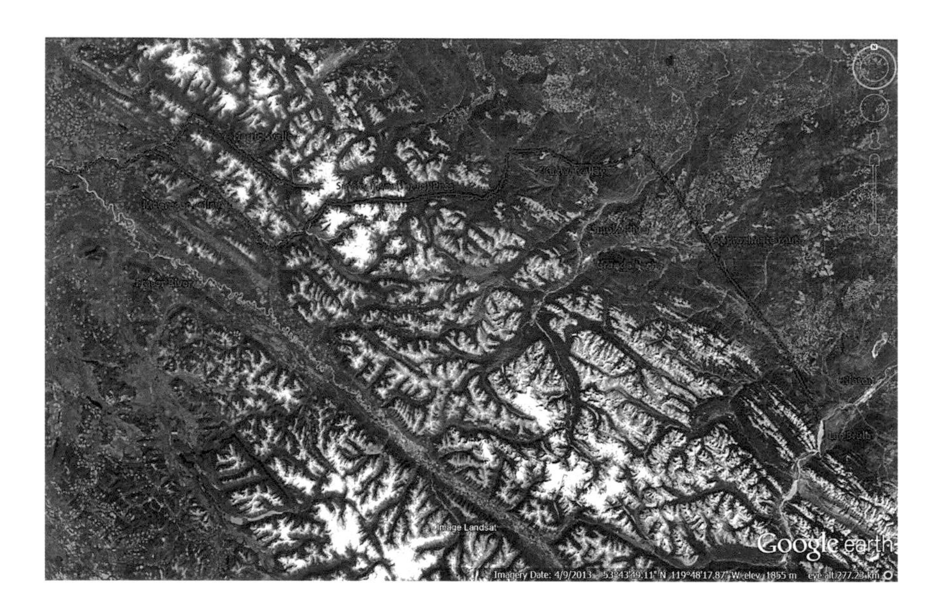

Figure 2. Jarvis and Hanington's route through the Rockies. IMAGE COURTESY OF GOOGLE EARTH

Figure 3. Jarvis and Hanington provided the first written descriptions of Herrick Falls. PHOTO COURTESY OF PAUL WALSH

Figure 4. Mount Ida. PHOTO COURTESY OF KEITH MONROE

Figure 5. Smoky River (Jarvis) Pass, looking east. IMAGE COURTESY OF GOOGLE EARTH

Figure 6. Kakwa Falls in winter; Jarvis and Hanington provided the first written descriptions. PHOTO COURTESY OF ALBERTA PARKS

Figure 7. The distinctive profile of Roche Miette. PHOTO COURTESY OF BRIAN CARNELL

Figure 8. Jarvis Lake (Alberta) in summer. This is one of the "three or four small lakes" that Jarvis referred to. PHOTO COURTESY OF BRIAN CARNELL

Figure 9. Brule Lake with Roche Miette in the distance. PHOTO COURTESY OF BRIAN CARNELL

Figure 10. Fiddle River, Jasper National Park, near the site of the depot. PHOTO COURTESY OF CHARLES HELM

Figure 11. The Clerk's Quarters, Fort Victoria, built in 1864, is now an Alberta heritage attraction. PHOTO COURTESY OF CHARLES HELM

Figure 12. Fort Carlton, restored and now a heritage attraction in Saskatchewan. PHOTO COURTESY OF CHARLES HELM

Figure 13. *Spathanaw watchi*, Round Hill, or Mount Carmel, now the site of a Roman Catholic shrine. PHOTO COURTESY OF CHARLES HELM

Figure 14. Council of Hudson's Bay Company commissioned officers held in Winnipeg, 1887. HUDSON'S BAY COMPANY ARCHIVES, ARCHIVES OF MANITOBA CUTHBERT AND MARY JANE SINCLAIR FONDS N9248. Three of the Hudson's Bay Company factors that Jarvis and Hanington encountered during their expedition appear in this photograph: Lawrence Clarke (Fort Carlton) is seated on the right in the first row; Richard Hardisty (Fort Edmonton) is seated behind him on the right in the second row; Archibald McDonald (Fort Ellice) is standing in the fourth row, second from right.

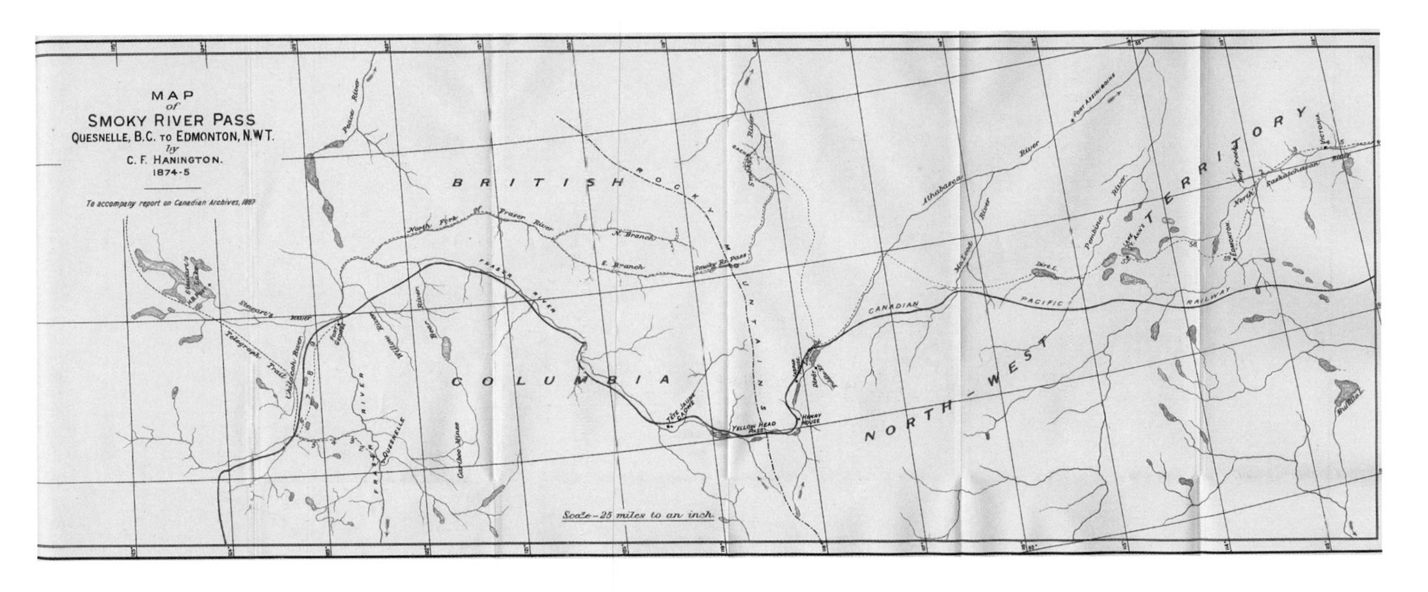

Figure 15. Hanington's map – portion to Edmonton. Reproduced by Brymner in "Report of Archivist", 1887, together with Hanington's journal.

Figure 16. Brown and Gillis's Occidental Hotel. QUESNEL AND DISTRICT MUSEUM AND ARCHIVES, P1958.302

Figure 17. Reid's Store. QUESNEL AND DISTRICT MUSEUM AND ARCHIVES, P1990.88

Figure 18. The Lac Ste. Anne pilgrimage site on the shores of the lake, at the site of the Roman Catholic Mission that Hanington visited in 1875. PHOTO COURTESY OF CHARLES HELM

Figure 19. South Fork of the McGregor River. Winter camp life as Jarvis and Hanington would have experienced it. COURTESY OF THE ROYAL BC MUSEUM, BC ARCHIVES, IMAGE F 03744

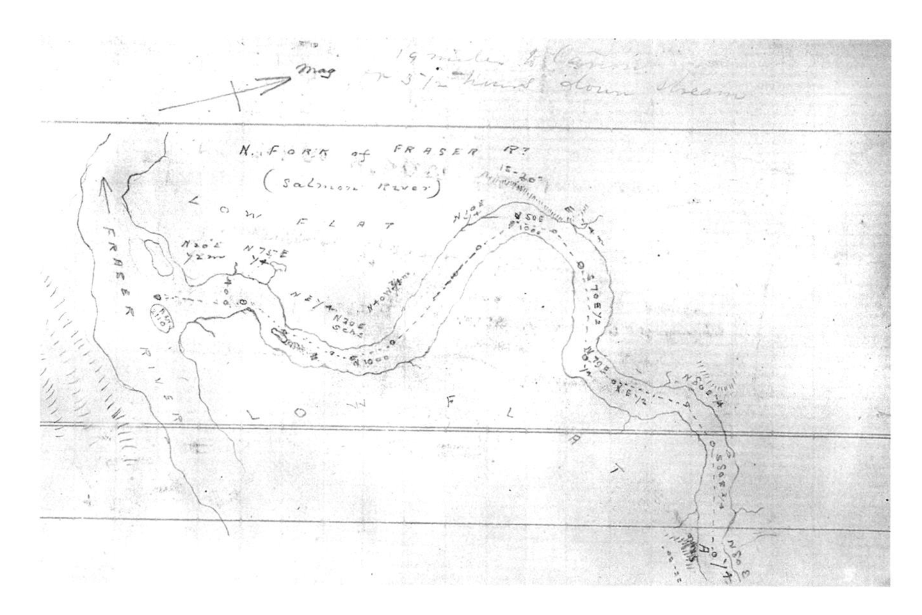

Figure 20. Jarvis's diary sketch of the confluence of the Fraser and McGregor Rivers.

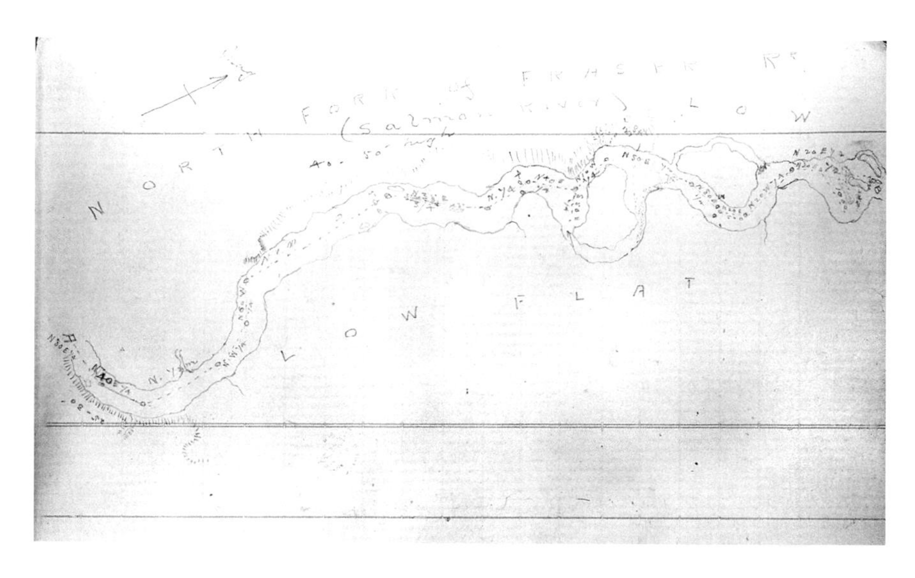

Figure 21. Jarvis's diary sketch of the lower McGregor River.

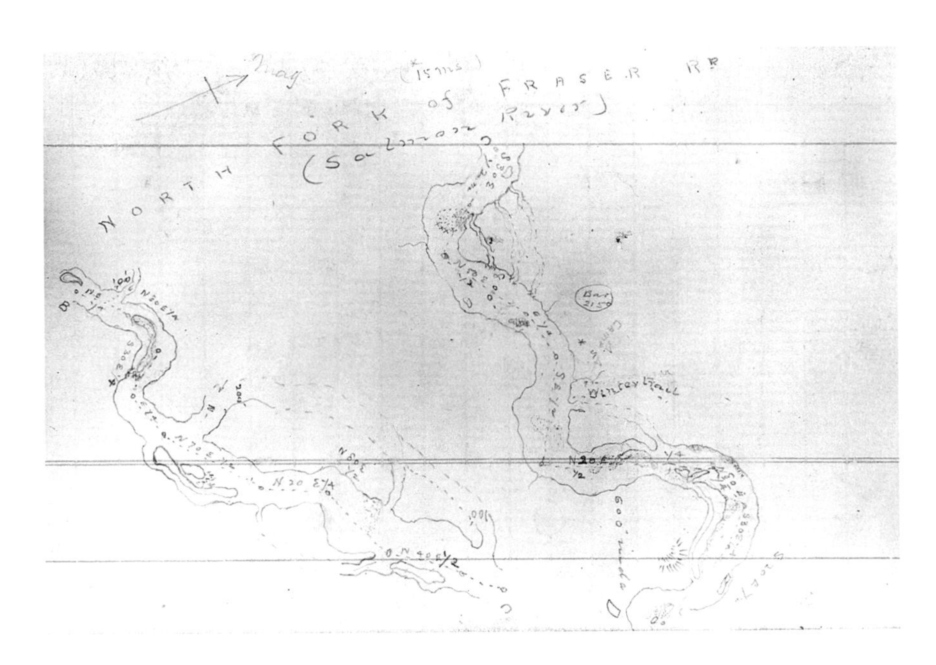

Figure 22. Jarvis's diary sketch of the lower McGregor River.

Figure 23. Jarvis's sketch of the first view of "Smoky River Pass."

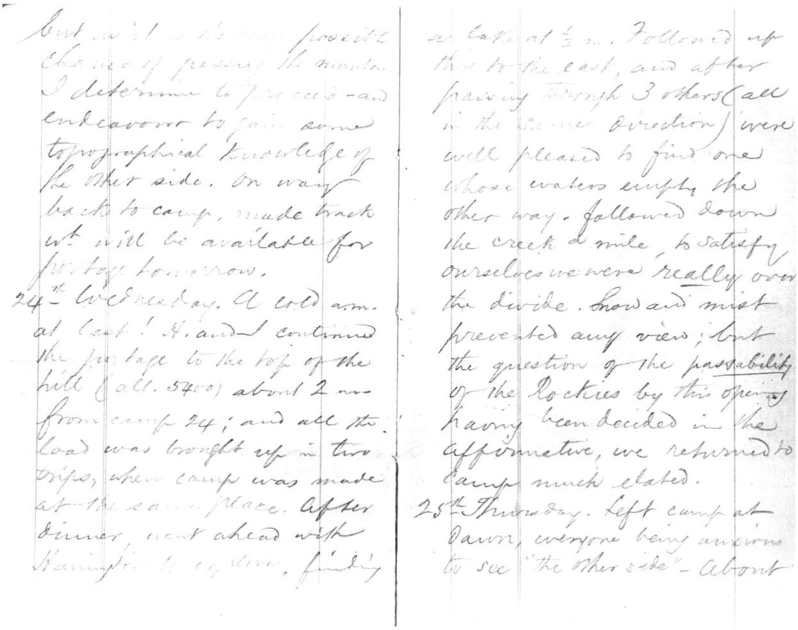

but as it is the only possible chance of passing the mountains I determine to proceed – and endeavour to gain some topographical knowledge of the other side. On way back to camp, made track wh. will be available for portage tomorrow.

24th Wednesday. A cold a.m. at last! H. and I continued the portage to the top of the hill (alt. 5400) about 2 m. from camp 24; and all the load was brought up in two trips, when camp was made at the same place. After dinner, went ahead with Hanington to explore, finding a lake at ½ m. Followed up this to the east, and after passing through 3 others (all in the same direction) were well pleased to find one whose waters empty the other way. followed down the creek a mile, to satisfy ourselves we were 'really' over the divide. Snow and mist prevented any view; but the question of the passability of the Rockies by this opening having been decided in the affirmative, we returned to camp much elated.

25th Thursday. Left camp at dawn, everyone being anxious to see "the other side" – About

Figure 24. Jarvis's diary entry for February 24, 1875.

Figure 25. Jasper House and Roche Miette, January 1872. NATIONAL ARCHIVES OF CANADA, MIKAN 3264788/PA-009147

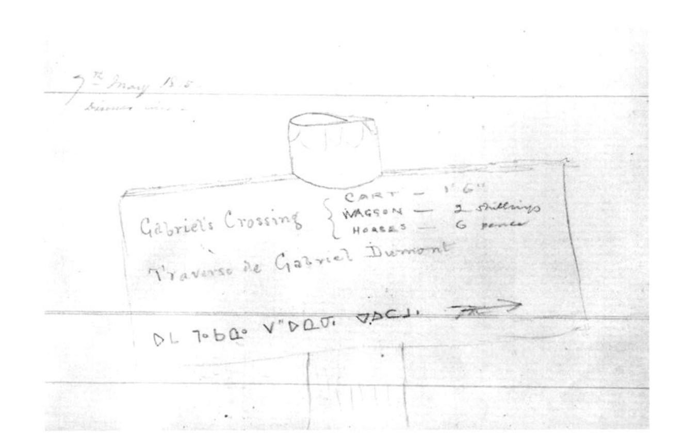

Figure 26. Jarvis's sketch of the sign near Mount Carmel, Saskatchewan, pointing to Gabriel Dumont's ferry crossing at the South Saskatchewan River.

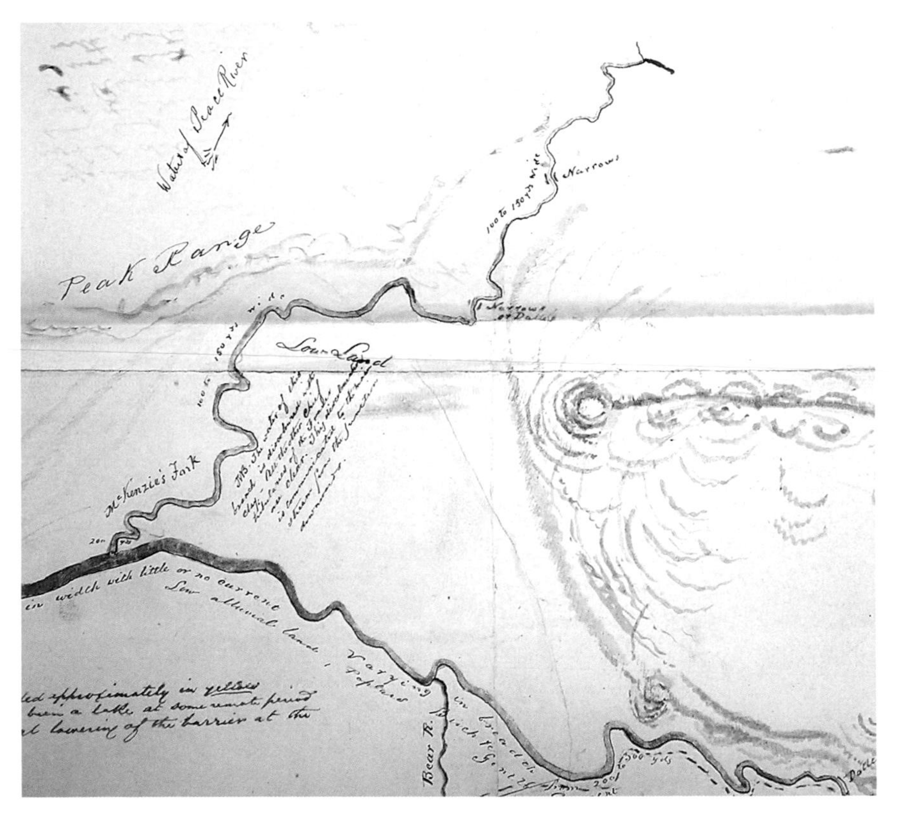

Figure 27. Part of A.C. Anderson's January 1874 map showing the confluence of the Fraser River and the McGregor River (labelled McKenzie's Fork). IMAGE CM/13699A, COURTESY OF THE ROYAL BC MUSEUM, BC ARCHIVES. EXTRACT FROM ORIGINAL MAP

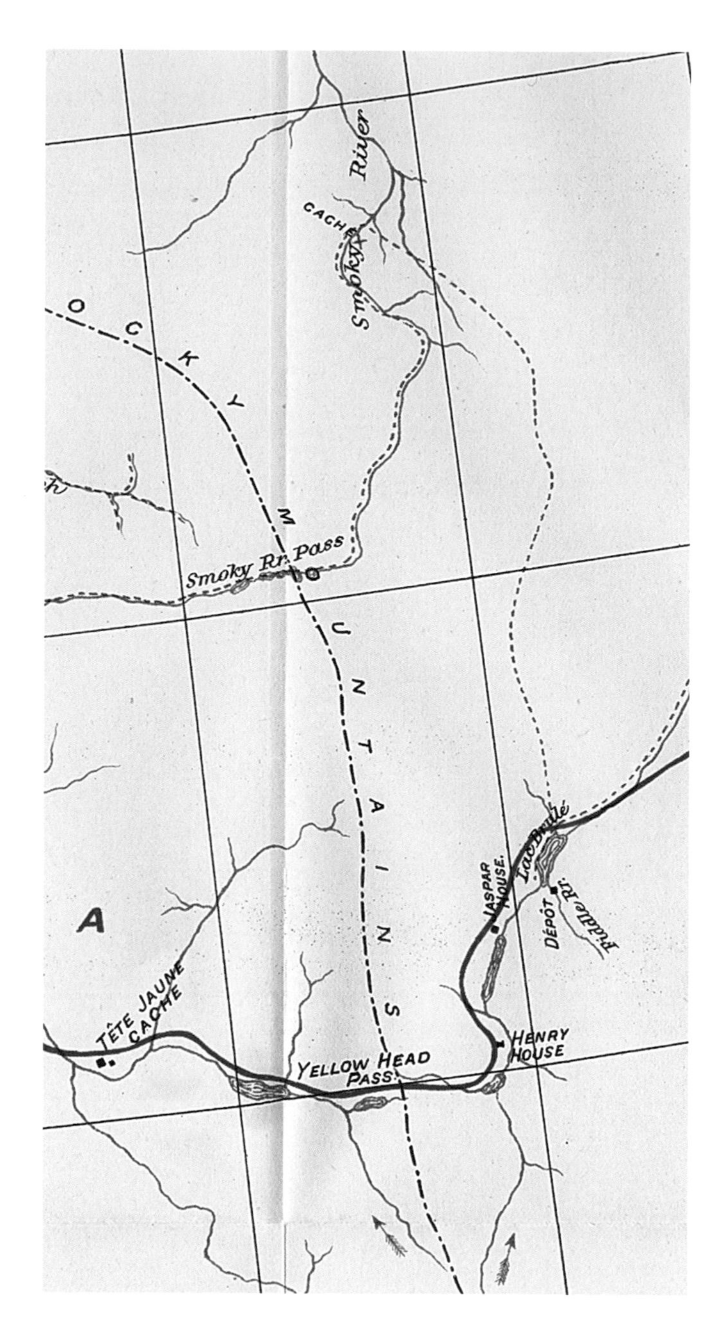

Figure 28. Hanington's map of the route they followed east of Smoky River (Jarvis) Pass.

Figure 29. E.W. Jarvis, Christmas 1872, two years before the Smoky River Pass expedition. PHOTO BY WILLIAM NOTMAN. NATIONAL GALLERY OF CANADA, ACCESSION # 34992.56

Figure 30. Mount Edward House in Charlottetown, the Jarvis family home where Jarvis was born and lived as a young child. PHOTO COURTESY OF CHARLES HELM

Figure 31. Jarvis and Berridge company boat *Ogema* on the Red River. GLENBOW ARCHIVES, IMAGE #NA-944-1

Figure 32. Winnipeg in 1880. GLENBOW ARCHIVES, IMAGE NA-1255-2

Figure 33. Jarvis and fellow North West Mounted Police officers in 1888. LIBRARY AND ARCHIVES CANADA, MIKAN 3711018/E008128862

Figure 34. Jarvis (back row, on right) and fellow North West Mounted Police officers in 1890. GLENBOW ARCHIVES, IMAGE PA-3886-3

Figure 35. Major E.W. Jarvis, "The Terror of the Horse Thieves," November 28, 1893. GLENBOW ARCHIVES, IMAGE NA-2307-8

Figure 36. North West Mounted Police barracks at Maple Creek in the 1890s.
GLENBOW ARCHIVES, IMAGE PA-3946-5-6

Figure 37. North West Mounted Police barracks at Fort Calgary in 1897.
GLENBOW ARCHIVES, IMAGE PD-383-1-9

Figure 38. Jarvis's grave in the Pioneer Section of St. Mary's Cemetery, Calgary. PHOTO COURTESY OF MIKE MURTHA

Figure 39. Mount Jarvis in centre, viewed from Jarvis Lakes. PHOTO COURTESY OF KEITH MONROE

Figure 40. Jarvis Creek (Alberta) flows out of Jarvis Lake in the William A. Switzer Provincial Park. PHOTO COURTESY OF CHARLES HELM

Figure 41. Mt. Jarvis from the south with Mt. Hanington in the background.
PHOTO COURTESY OF BILL AND BEV RAMEY, PARK HOSTS, KAKWA PROVINCIAL PARK

Figure 42. C.F. Hanington in 1878, three years after the Smoky River Pass expedition. PHOTO BY WILLIAM J. TOPLEY. LIBRARY AND ARCHIVES CANADA, MIKAN 3446842/E010971282

Figure 43. Crowsnest Pass railway tunnel alongside Moyie Lake, in the section under Hanington's supervision. GLENBOW ARCHIVES, IMAGE ND-9-18

Figure 44. C.F. Hanington in the late 1890s. COURTESY OF THE ROYAL BC MUSEUM, BC ARCHIVES, IMAGE G-07580.

Figure 45. The Thompson River Canyon near Spence's Bridge. COURTESY OF THE ROYAL BC MUSEUM, BC ARCHIVES, IMAGE H-01328

Figure 46. Canadian National Railway construction survey party in the Thompson River Canyon, the section under Hanington's supervision.
KAMLOOPS MUSEUM AND ARCHIVES PHOTO COLLECTION, #394

Figure 47. Major Charles Francis Hanington in 1917. PHOTO COURTESY OF DAVE MACLENNAN

Figure 48. Emma and Frank Hanington. PHOTO COURTESY OF DAVE MACLENNAN

Figure 49. Mount Hanington viewed from alpine meadows south of Jarvis Pass. PHOTO COURTESY OF BILL AND BEV RAMEY, PARK HOSTS, KAKWA PROVINCIAL PARK

Figure 50. Kakwa Falls. PHOTO COURTESY OF CHARLES HELM

Figure 51. Kakwa Falls from within the cave. PHOTO COURTESY OF CHARLES HELM

Figure 52. Samuel Prescott Fay's photograph of Mount Ida from "Matterhorn Camp." PHOTO COURTESY OF *CANADIAN ALPINE JOURNAL*

CHAPTER 5

The Expedition in Perspective

Starting Out

When Jarvis and Hanington set out from Prince George in mid-January 1875, they were not on a vague search for a putative pass through the Rockies. The existence of Smoky River Pass was known, as was its approximate location, but only through word of mouth. Their mission was to find the pass, pinpoint its location and determine its suitability for railway purposes.

Aboriginal hunters and trappers had been crossing the Rockies for countless generations, using a variety of passes. Jarvis himself noted the evidence of such use in Smoky River Pass and the McGregor River valley. The arrival of the fur trade in central British Columbia in the early 1800s probably intensified trade across the mountains. In 1819, two years before the amalgamation of the Northwest Company and the Hudson's Bay Company, HBC Factor Colin Roberts sent a party out from Fort Dunvegan, near the confluence of the Smoky River and the Peace River, to cross the Rockies and draw trade away from the Northwest Company. The party made their way up the Smoky River, cached supplies at what is now Grande Cache and probably crossed the mountains via Sheep Pass, a little south of Smoky River Pass (now Jarvis Pass).[1] In 1835, HBC trader A.C. Anderson canoed up the Fraser River from Fort (Prince) George to Tête Jaune Cache and his Iroquois guide pointed out a transmountain route via Rivière à la Boucane (now Torpy River) that he said connected with Smoky River. Anderson shared this information and a map that he drew with Marcus Smith in May or June 1873,[2] and this is likely what prompted Jarvis's and Hanington's expedition (see Figure 27).

One of Jarvis's Canadian Pacific Survey colleagues was Walter Moberly, who explored extensively in the Jasper area in 1872 and 1873 (and who built the Fiddle River depot that Jarvis and Hanington headed for). Moberly's brother John had been an HBC factor at Jasper House in the 1860s and hunters from the post regularly travelled north into the Smoky River watershed.[3] They would have known of the passes through the mountains. John Moberly probably passed this information along to his brother Walter, who would have shared it with Fleming. Walter Moberly himself, in *The Rocks and Rivers of British Columbia*, mentions that one of his native assistants, Louis, had "gone after his people to the Smoky River."[4] Alfred Selwyn of the Geological Survey of Canada explored the Peace River in the vicinity of Smoky River in 1875 and noted that Aboriginal people travelled to the area each summer via a good horse trail from Jasper House, taking about ten days for the trip.[5]

The 1832 map by London cartographer John Arrowsmith showed both the North Fork of the Fraser River and Smoky River in rudimentary form, in approximately the correct location; the same information appeared on the 1854 Arrowsmith map. Similar information was also presented in the 1871 map of British Columbia, commissioned by the Honourable J. Trutch, Commissioner of Lands.[6]

By January 1874, then, the CPR Survey had information about both the upper Fraser and the Smoky rivers, which appeared on Sandford Fleming's map. This is the map that Jarvis and Hanington used.[7] Additionally, Marcus Smith and Jarvis spent a few days together in Fort (Prince) George in the fall of 1874, just prior to the expedition. Undoubtedly, Smith shared the information he had received from Anderson.[8]

Jarvis and Hanington started out with Fleming's map and the information via Anderson of a comparatively short crossing over the mountains. Both were misleading. However, Jarvis was experienced enough to plan for a much longer trip and he was known for his organizational skills, as noted by Smith in 1873. He had Hanington canoe up the Fraser and McGregor rivers in the fall of 1874 to cache supplies in advance. He arranged for dried salmon to be obtained from Quaw's cache at Bear (Bowron) River. He kept equipment to a minimum to lighten loads (and, consequently, there are no photographs) and he planned enough supplies for two months. He was well prepared. In later testimony to the Royal Commission investigating the CPR, he stated that supplies would have been adequate for the expedition as envisaged.[9] However, the expedition was plagued by several problems beyond his control – the fruitless two weeks spent searching along the Herrick Creek valley, the rough terrain that necessitated numerous portages around canyons and the very changeable weather that slowed travel. Together, they turned the expedition into a two-and-a-half-month ordeal. Extremely difficult travel conditions, for a much longer period, exhausted men and dogs and the food supply dwindled to nothing.

Once the party reached the summit of Smoky River Pass, Jarvis had to make a difficult decision. He and Hanington had achieved their primary objectives. They had found the pass, determined its geographic coordinates and altitude and established its unsuitability for a railway line. Food was running out and they were unsure of their location in relation to the Athabasca valley. They could return to Fort George, but it would be a long, hard trip and their outbound tracks were probably snowed over. Alternatively, they could continue to follow approximately the "unexplored route" on their map to the Athabasca River valley where they would find supplies at Jasper House. This choice would also provide more information about the blank area on the map and, as a dedicated employee, Jarvis wanted to complete his assignment. With Hanington's urging, he chose to continue.[10]

Years later, Hanington noted in his reminiscences that they had been assured that they "would find supplies of all sorts" at the Fiddle River depot in the Athabasca valley. After crossing the pass, they headed towards this depot and the post at Jasper House, unsure of the route, feeling their way over mountain spurs and river valleys, in a desperate race against starvation. However, they found the depot devoid of supplies and the post abandoned. Hanington commented that "some official of the government had ordered those supplies sent to Edmonton, and there was nothing left for hungry men."

They were saved only by the generosity of some local Native families who shared their own meagre food supplies with the strangers. The party, at the limits of their endurance, finally stumbled into Lac Ste. Anne, 100 km northwest of Edmonton, 13 days later. They spent several days recovering in Edmonton but suffered severe gastro-intestinal symptoms following their previous starvation. Then it was on to Winnipeg, their end point, by horseback, where they arrived almost seven weeks later, on May 21. The whole trip took five and a half months.

Where Is Smoky River Pass?

There is no Smoky River Pass on modern maps. Jarvis was directed to explore the North Branch of the North Fork of the Fraser River, which is now known as Herrick Creek. There is no pass to the east at the head of this valley, but he did accurately plot the latitude at 54°08' before turning back. The pass that Jarvis and Hanington subsequently crossed is now known as Jarvis Pass. There are two other passes in the vicinity, McGregor Pass and Sheep Pass, either of which could be the putative Smoky River Pass. Sheep Pass has a gentler gradient and is wider than the other two; it is likely the pass that Anderson's guide alluded to when he pointed to the Rivière à la Boucane (which translates as Smoky River, i.e., connecting with the Smoky River). A short paddle up this river, now the Torpy River, would allow a short portage to the upper McGregor River, which then gives access to Sheep Pass via the Bastille Creek valley. To add to the confusion, the next Fraser River tributary east of Torpy River is the Morkill River, which in the past was known variously as the Little Smoky River and the Big Smoky River.[11] This river and its tributaries also lead to passes across the Rockies. In all likelihood, each pass was used at times by Aboriginal people – they are all connected by large expanses of alpine terrain where there is comparatively easy travel, plus, in the past, large caribou herds and, according to Anderson's guide Réné, bison also.

All of these alternative passes are further south and would not have attracted Jarvis and Hanington's attention even if they had been aware of them. When they reported that they turned northeast and entered the mountains, they were following a 90° turn of the McGregor River. They were at 53° 55'N and following a map that indicated that they should cross a pass in the vicinity of 54° 30'N. The valley then trends northeast in a straight line for about 27 kilometres until turning east towards the pass – exactly where they needed to go. Halfway along, they would have crossed a tributary valley that is now considered the headwaters of the McGregor River (and the route to Sheep Pass), but it enters from the southeast, following the base of the high mountains. They rightly chose to ignore it.

Their initial excitement at seeing the pass quickly turned to disappointment when they encountered a steep western approach and a crossing estimated at 5,400 feet, two conditions that eliminated it from further consideration, especially compared to Yellowhead Pass. Modern maps show the elevation of the pass as 4,900 feet. The Fleming map indicated the pass at 54° 30'N, 119'W. Hanington's map shows the pass at 54° 05'N, 118° 35'W, and modern maps indicate 54° 09'N, 120° 12'W. They were very close in their latitude calculation and longitude was always more difficult to determine; there was a considerable error in the map they were following. At that latitude, each degree of longitude measures 40.75 miles (65.5 kilometres). In reality, both the Fleming map of 1874 and Hanington's map, prepared in 1875, used similar grids, so the longitude errors were inconsequential, especially for their reconnaissance purposes (see Figure 28).

Why a Winter Expedition?

Looking back 140 years, the decision to conduct a hazardous winter expedition appears puzzling. Even at the time, the dangers of winter travel through the mountains were known. In his reminiscences, Hanington wrote that the experienced and bush-wise miners in Quesnel "had a very exalted idea of the pleasure to be derived from our trip across the mountains and we heard many prophecies in regard to our going to destruction. In fact the last words we heard were 'God bless you old fellows' – goodbye; this is the last time we will see you."[12]

Sandford Fleming did not order or request a winter survey. In later testimony before the Canadian Pacific Railway Royal Commission, Jarvis made it clear that it was his decision: "The Chief Engineer wished an exploration to be made in the mountains, and I volunteered to make it during the winter. He would not issue any instructions to that effect, but he simply said he wished another exploration made north of Tête Jaune Cache, through the Rocky Mountains."[13]

Jarvis was a dedicated employee of the Canadian Pacific Survey and had great respect for Fleming, for whom he had worked since 1867. They were personally acquainted and Jarvis had attended social events at the Flemings'.[14] Undoubtedly, Jarvis knew of the urgency to conclude route surveys so that construction could begin; the commitment to complete the railway by 1881 was looming. It is likely that, having been selected to undertake the survey, Jarvis was anxious to complete it as expeditiously as possible.

Jarvis and Hanington were no strangers to winter surveying. Jarvis had worked through the winter for the Intercolonial Railway and they had both spent winters outdoors in northwestern Ontario. They were experienced with dogsleds and snowshoes. And there were precedents in the Rockies – the Hudson's Bay Company had been making winter crossings of Athabasca Pass for many years and Jarvis's colleague Walter Moberly had been active in the Jasper area during the previous two winters. Both men had been in the region for a couple of seasons and had some familiarity with both the Rockies and the nearby Cariboo Mountains, including high altitude crossings of glaciers in early spring.[15] There were likely practical considerations as well. From their travels along the upper Fraser Valley, they would have been familiar with the dense interior rainforest on the western slopes of the Rockies where they would be travelling. Along the Fraser valley, transportation on the river made travel a little easier, but in the heart of the mountains there would be no such opportunity. Winter travel along the open terrain of frozen rivers made a lot of sense.

They were not naive about the undertaking. They were very careful and thorough in their preparations – exercising to keep fit, recruiting dog teams and Aboriginal assistants, ensuring adequate supplies for the anticipated length of the trip and waiting for the rivers to freeze and create the right travel conditions. Everything suggests they understood the challenges, were physically and mentally ready and set out well organized and equipped. As a result, their expedition achieved its objectives, even though unanticipated delays tested them all to the limit. Their daring saved Fleming a whole field season.

E.W. Jarvis and C.F. Hanington's Accomplishments

In his book *The National Dream: The Great Railway, 1871–1881*, Pierre Berton devotes a few pages to the Jarvis-Hanington expedition and concludes with the question about all the engineers and surveyors:

> Why did they do it? Why did any of them do it? Not for profit, certainly, there was little enough of that; nor for adventure, there was too much of that. The answer seems clear from their actions and their words: each man did it for glory, spurred on by the slender but ever-present hope that someday his name would be enshrined on a mountain peak or a river or an inlet, or – glory of glories – would go into the history books as the one who had bested all others and located the route for the great railway.[16]

In the case of Jarvis and Hanington, this is an uncharitable characterization of their motives. It was certainly a memorable adventure and Hanington was still bubbling with enthusiasm 50 years later. Their devotion to Fleming was undoubtedly a factor. And they would have been pleased had they found the ideal route, as would anyone involved in the great enterprise. But nothing in their subsequent lives hinted at any aspiration for fame and glory. Neither made any attempt to publicize or profit from the expedition. Jarvis immediately wrote his low-key official report, then left the survey for private business. Hanington simply returned to his life as a route surveyor. His own journal was written for the pleasure of his brother and had not been intended for publication. It was donated to the National Archives without his knowledge. Their reports appeared, 11 years apart, in obscure government reports, published by others, with no thought of a broad public audience. To this day, they remain known only to a few enthusiasts interested in railway history or early exploration in the northern Rockies.

Both were modest men. The *Manitoba Free Press* obituary for Jarvis noted "Major Jarvis was not a man who had ever occupied a large space in the public view," despite his many contributions to the early development of Winnipeg.[17] Hanington's obituary was equally low key: "Last of a notable group of engineers who were associated with Sir Sandford Fleming in surveying proposed routes for the trans-continental line of the Canadian Pacific Railway, Major Charles Francis Hanington...spent much of his life in exploration and surveys in the far West."[18]

A more likely explanation of their willingness to undertake the expedition is a combination of youthful enthusiasm and patriotism. They had an opportunity for adventure, and, as unmarried young men, they grabbed it. It was part of a grand nation-building enterprise, an opportunity that would never come again. Canada was only seven years old and the transcontinental railway was the first grand vision of the new nation. The period was also the high point of British imperial expansion and Jarvis and Hanington were undoubtedly imbued with Victorian ideals and the obligation of service. Hanington spoke boldly of his admiration for the Empire when he enlisted in the First World War at the age of 68. Jarvis had been raised and educated in England, probably at a public school. He was later a militia volunteer who served during the Northwest Rebellion, then joined the recently formed North West Mounted Police, another nation-building organization. Both men were from well-connected Maritimes families but found life on the western frontier very congenial and enthusiastically participated in the opening of the west.

In the big picture, their contributions were modest, but, in combination with those of many others, they collectively contributed to the success of the major undertaking that was the Canadian Pacific Railway. They also helped fill in one of the blanks on the map of British Columbia, and their information was quickly incorporated by George Dawson in his map of 1879.[19] Their daring expedition helped narrow the range of choices for crossing the Rockies. The expedition itself was a major achievement and must have given them lifelong satisfaction in a job well done. Certainly, later travellers were full of admiration and Andrews is correct in recognizing it as an epic of Canadian exploration.[20] Their "Chief," Sandford Fleming, in his 1877 summary report, recognized their "hazardous expedition" and noted that "they suffered unusual hardship."[21] Fifteen years later, he chose to single out their expedition in a 1889 paper to the Royal Society of Canada about "Expeditions to the Pacific."[22] Of the many expeditions undertaken by Canadian Pacific Survey members, this was the one that he highlighted as noteworthy.

Remarkably, both men were only in their 20s at the time, but Fleming obviously had great faith in their abilities. They overcame enormous physical and mental challenges to safely complete the work. Even after reaching Smoky River Pass, when the logical decision would have been to return to Fort George, they determined to complete their assignment – a tribute to their loyalty, stamina, resolve and abilities. Fleming's trust in them was well founded.

Jarvis summarized their satisfaction in his own words as they approached Winnipeg: "We were just reaching the goal, pushed forward over many a weary mile of mountain and plain, and could take our well-earned repose in the happy consciousness of having fulfilled the task allotted to us, and earned the approbation of him we are proud to acknowledge our Chief."[23]

The "task" that Jarvis referred to so unassumingly, an expedition from Quesnel to Winnipeg, involved 165 days and 106 camps; 3000 kilometres of travel, of which 1600 kilometres were on snowshoes and 630 kilometres were through unknown terrain; 20 consecutive days of -30°C temperatures; two blankets for a bed and a piece of cotton for a shelter; short rations and, finally, no food. And through it all they kept counting their paces for the survey – all 2,188,900 of them. Truly an epic. Jarvis and Hanington had written themselves into the annals of Canadian history, as the leaders of one of the most remarkable expeditions in the young country's evolution.

CHAPTER 6

BIOGRAPHIES

Major Edward Worrell Jarvis

Edward Worrell Jarvis (see Figure 29) was a member of a prominent Prince Edward Island family that was descended from Loyalists who had left Connecticut.[1] As a young man he witnessed the birth of Canada in 1867 and the entry of his home province into Confederation in 1873. In his comparatively short life he had a varied career and the good fortune to participate in a number of seminal events in Canadian history. He wrote an official account of the exploration that is the focus of this book. He also kept diaries and field notebooks, several of which have fortunately survived and are preserved at the Archives of Manitoba.[2] Most notable are the 1875 notes and sketches of the expedition through Smoky River Pass. As far as we are aware, the diaries and notebooks were not previously known and have not been published, apart from their use in an obscure 1956 young adults' novel based on the diaries.[3]

Jarvis was born in Charlottetown (see Figure 30) on January 26, 1846.[4] His father, the Honourable Edward James Jarvis, was Chief Justice of the then colony of Prince Edward Island, and his uncle, Colonel John Hamilton Gray, was premier of Prince Edward Island from 1863 to 1865, during which time he chaired the Charlottetown Conference and became a Father of Confederation. His mother, Elizabeth Gray, was his father's second wife. Edward James Jarvis had eight children from his first marriage and Edward Worrell Jarvis was the second of three children from the second marriage. Five of his siblings died before his birth.[5]

When Jarvis was less than two years old, his mother died, in September

1847, during the birth of her third child, and he became an orphan at age 6, when his father died in May 1852.[6] His sister Amelia, then 17, raised him for a short while and he was then sent to England in September of 1852 for his education,[7] possibly to relatives, as his will mentions a cousin in Birkenhead, Cheshire, and as a teenager he travelled in Europe with an English aunt.[8] He obtained a degree in civil engineering at Cambridge University, and served a three-year pupilage with the Great Northern Railway in England.[9] (He later was elected an associate member of the UK Institution of Civil Engineers in May 1874.)[10]

In late 1867, he returned to Canada to work for Sandford Fleming as a surveyor and engineer on the Intercolonial Railway in Nova Scotia, which had been a condition of the province's entry into Confederation.[11] He worked on route surveys and locations along the Northumberland shore and in the Wentworth area, and designed the Smith Brook embankment, which, at the time, was the largest in the Dominion.[12]

When Fleming took over responsibility for the Canadian Pacific Railway Survey, Jarvis followed. In 1871, he began work as an engineer in charge on location surveys in northwest Ontario, until June 1873, when he was transferred to British Columbia.[13] The remainder of 1873 and 1874 were spent mostly in the interior of British Columbia, exploring potential routes between Lillooet and Yellowhead Pass, across the Cariboo Plateau, through the rugged Cariboo Mountains and along the North Thompson Valley.[14] It was in 1871, in northwest Ontario, that he first met and worked with Hanington, who had been appointed as a rodman, and they worked together during the next three years.[15] The pair obviously impressed Fleming, as he chose them for the perilous winter exploration of Smoky River Pass. Jarvis, with his experience as an engineer in charge, was given leadership.

Following the successful completion of the project, Jarvis and Hanington travelled to their headquarters in Winnipeg, where they prepared their report for Fleming. Both were invited to return to British Columbia as location surveyors, but Jarvis chose to resign from the CPR Survey, noting in his diary on June 7, 1875: "obliged to refuse as I can't stand any more B.C."[16] The Smoky River expedition had tested him to the limit, especially as he had the responsibility for the safe return of the entire crew – and there were times when he must have wondered whether they would make it. Hanington mentions in a journal entry dated May 4, 1876, that Jarvis "said once that the mention of Smoky River made him shudder, and I dare say it would."[17]

Jarvis stayed in Winnipeg, settled into the community, and immediately entered the lumber business with George Macaulay. Diary entries for the first half of 1873 indicate that the partnership had developed in that earlier period.[18] Winnipeg was booming because of the anticipated arrival of the CPR and the opening of the west (the city grew from 3,000 residents in 1875 to 8,000 by 1880[19]), and the "Macaulay and Jarvis" company was a success. By 1877, they were importing 4.5 million logs and 30 train car loads of lumber from Red Lake, Minnesota.[20] On January 1, 1880, the company was succeeded by "Jarvis and Berridge," also a lumber company, that expanded to three sawmills and 50 – 150 employees seasonally. They had their own steam tug, the *Ogema* (see Figure 31), to tow logs on the rivers and lakes.[21] The business suffered a setback when the boiler exploded in May 1880, causing at least one fatality and destroying much of the mill.[22] However, the mill was still operating in 1883 when timber was being supplied from Rat Portage (now Kenora) in northwestern Ontario.[23]

During his time in Winnipeg (see Figure 32), Jarvis was an active member of the community. In 1875, during his first year in the city, he had a house built in Point Douglas.[24] He ran for public office several times and served as an alderman in 1876. Also in 1876, his name was put forward for mayor, but he declined to run. He was a founding director of the Winnipeg Curling Club and the Manitoba Historical Society. In December of 1877, he was appointed the first registrar of the newly founded University of Manitoba. Also during 1877, the Winnipeg City Council appointed him to a committee that was formed to encourage and aid the construction of a railway bridge at a suitable location over the Red River, to counter the proposed routing of the CPR line further north, via Selkirk.[25] He served as a member of the Roman Catholic section of the provincial board of education.[26] In addition, *Henderson's Directory of the City of Winnipeg* for 1880 lists him as a 1st lieutenant in the artillery of Military District 10; a council member of the board of trade and of the Manitoba Rifle Association; vice-president of the Winnipeg Cricket Club; a director of the Winnipeg General

Hospital and a member of the Committee of Management of the Manitoba Club.[27]

Jarvis put his education as a civil engineer to good use by designing the first bridge over the Red River, the Broadway Bridge, which linked Winnipeg and St. Boniface. It opened in April 1882. He also designed the Louise Bridge across the Red River and the Main Street Bridge across the Assiniboine River.[28]

Jarvis became a major in the Winnipeg Field Battery, a volunteer militia. During the 1885 Northwest Rebellion, the battery was assigned to the North West Field Force and fought at Batoche as part of Colonel Middleton's column in the main engagement in May.[29] Following the defeat of Louis Riel and the Métis and Aboriginal resistance on May 12, one of Jarvis's subordinates handed him Riel's diary of the previous few weeks, which had been recovered from one of the buildings. Jarvis took the diary and other papers to the nearby town of Prince Albert, the military staging area, for forwarding to Ottawa. On the second day of the subsequent trial of Louis Riel, on July 29, 1885, Jarvis testified as to the authenticity of the papers.[30] The diary later remained with Jarvis when he returned to Winnipeg. It was known at the time that he had the diary, but following his death in 1894, this important piece of Canadian history was presumed to be lost.[31]

However, the Riel diary was unexpectedly found in Winnipeg, together with Jarvis's personal diaries, in the early 1950s.[32] They were discovered in an old desk that had belonged to the Alloway and Champion bank. The bank's founders, William F. Alloway and Henry T. Champion, may have had a business relationship with Jarvis when he was a partner in the lumber businesses. Certainly, there was a personal relationship, as the two bankers were the executors of Jarvis's will and one of his bequests was to Champion's young daughter, Evelyne.[33] The University of Saskatchewan Library eventually acquired the Riel diary, and the Jarvis diaries were deposited at the Archives of Manitoba.[34]

Jarvis conducted himself well as a field commander at Batoche (he was mentioned in dispatches[35]), and a year later he made another dramatic career change. On April 10, 1886, he received a commission as a superintendent in the North West Mounted Police.[36] His regimental number as an officer was O.73. He was assigned to the command of "B" Division, headquartered in Regina and responsible for southeast Saskatchewan (see Figures 33–37).[37] His medical examination provides a physical description: "5' 8" and 185 lbs, with blue eyes, light brown hair, and slightly bald." His temperament was described as "sanguine" and his intelligence as "remarkably good."[38]

In late 1890, Jarvis was transferred to Maple Creek in southwest Saskatchewan, where he assumed command of "A" Division, whose area of responsibility extended as far as Medicine Hat in Alberta.[39] He stayed in this position until the end of 1892.[40]

Jarvis was transferred again in early 1893 to the command of "E" Division in Calgary.[41] Calgary was then a very young and bustling boomtown. The North West Mounted Police post of Fort Calgary had been established less than 20 years earlier, in 1875. The town took off with the arrival of the CPR in 1883 and was incorporated in 1884. The next ten years saw enough growth for Calgary to become a city on January 1, 1894.[42] As a leading government representative, Jarvis was a prominent member of the community. He was a founding member of the "Pack of Western Wolves," a men's social club that included prominent citizens such as lawyer James Lougheed and rancher Bill Cochrane. The club continues in existence as the Ranchmen's Club.[43]

Life was going well for Jarvis. He was a prominent and active member of a thriving new community that had boomed with the arrival of the Canadian Pacific Railway – the railway line that he had helped to locate during his survey work 20 years earlier. He was rising in the ranks of the North West Mounted Police and was a well-liked and well-respected commanding officer.[44] Then, unexpectedly, tragedy struck. He had visited Winnipeg in the fall of 1894 and was in good health, but in November he developed increasing swelling in the neck and respiratory distress, apparently following a sore throat.[45] Several doctors attended, but in pre-antibiotic days, treatment options were more limited than they are today. Jarvis lingered for about 12 days and died on November 26, 1894, just two months before his 49th birthday. "Cellulitis" was named as the cause of death.[46] North West Mounted Police members travelled on horseback from as far as Fort MacLeod to attend his funeral, which was conducted by a visiting

Catholic bishop, and flags flew at half-mast in Calgary.[47] Jarvis was buried in the Pioneer Section of St. Mary's Cemetery in Calgary (see Figure 38).[48]

As far as is known, Jarvis never married. No record of a marriage has been found; his cousin Florence Gray's family history notes record him as "died unmarried," and in his will he bequeathed his estate to family and friends, especially his nephews and nieces, who were the children of his brother Henry and his sister Amelia.[49] And yet his diary for the latter part of 1875 contains some tantalizing entries. One of his first commitments after starting in the lumber business was to have a house constructed in fashionable Point Douglas – an unusual priority for a busy bachelor. On August 11, he wrote: "Jour de bonheur, moving some furniture into the house – et touchant d'autres choses."[50] In his diary for August 24, he noted "Slept at house for first time."[51] On September 2, he wrote "Katie came today and we took our first meal at the house," and on September 12, "dined with Provencher and he took tea with us."[52] Both entries are suggestive of a partner, and in those times and with his social position, that could only have been a wife. The other puzzling clue is the 1881 Census, in which Jarvis's marriage status is listed as "widowed."[53] The mystery has not been solved.

Jarvis's contributions to the early development of Winnipeg are commemorated by the name of Jarvis Street and the historical information plaque at Provencher bridge.[54] Further west, his epic exploration of Smoky River Pass is well commemorated by numerous geographical names. Smoky River Pass is now Jarvis Pass and the summit lakes are Jarvis Lakes. The westernmost lakes form the headwaters of Jarvis Creek. Mount Jarvis overlooks the east end of the pass (see Figures 39–41). All of these features are in Kakwa Provincial Park in British Columbia.[55] A little further east are Jarvis Lake and Jarvis Creek in Alberta's William A. Switzer Provincial Park, and Jarvis Street, in the town of Hinton.[56]

Jarvis was a fascinating and complex person. A scion of a prominent, upper middle class, colonial family, he was raised and educated in England from the age of 6 and a half, almost certainly at a public school (i.e., an upper-class private school). In 1863–1864, when he was 17 and 18, he went on an extended European tour with "Dear Aunt" Agatha, spending at least eight months in Germany, primarily in Dresden (his diary after March 1864 is lost). The diary is full of descriptions of museums, palaces, art galleries and theatres, typical of the grand tours of the period by the affluent. Following the tour, he attended Cambridge University and graduated with a degree in civil engineering. In his diaries, he frequently mentions socializing with leading citizens in various communities. He was a member of the Carleton Club in Toronto, appreciated good hotels and fine dining and made holiday visits to Boston, New York and Europe.[57] He had a wide circle of friends and acquaintances and kept up a regular correspondence with them, including his colleague Hanington.[58]

And yet, after growing up in a seemingly privileged and cultured lifestyle in England, he was equally at home in the Canadian wilderness. From the fall of 1867 until summer 1875, while conducting railway location surveys, he spent most of his time camping in primitive conditions, much of it during long, cold winters in places such as northwestern Ontario. Travel was mostly by snowshoe and dog team (good preparation for his later Smoky River Pass expedition) and by canoe, on foot and on horseback in the summers. Meals consisted mostly of the usual "bacon, beans and bannock." Later, as a superintendent in the North West Mounted Police, he made frequent horse patrols, summer and winter, in the southern prairies. He was at home in small, rough-edged, frontier communities such as Winnipeg, Maple Creek and Calgary and, as a prominent, well-educated and gregarious citizen, played an active role wherever he went. He loved a good party and dance and his 1873 diary suggests he was often the instigator of such social events in Winnipeg (the population at the time was about 3,000).

Jarvis was also well-read and cultured – his diaries mention books that he was reading (e.g., August 28, 1863: "reading John Locke on Education" [a philosophical treatise from 1693]), quote excerpts from poems, list books and music to search out and record theatre and opera performances he had attended. He spoke French and German and made occasional cryptic notes in both, and during his time in Germany he also started learning Russian, Spanish and Italian. In his account of the expedition, he quotes from Virgil in Latin, likely learned at school in England. While working in northwest Ontario, he also learned some Ojibwa from one of his Aboriginal guides.[59]

Jarvis was an active churchgoer, frequently attending morning and evening services on Sundays. His family was Anglican and he was baptized at St. Paul's Anglican Church in Charlottetown, PEI.[60] However, at some point he converted to Roman Catholicism and during his time in Winnipeg he was a member of the Catholic section of the Manitoba Provincial Board of Education.[61] He appears to have been ecumenical, as his diaries mention occasional attendance at Anglican and Unitarian churches and he maintained a good relationship with his older brother Henry who was a prominent Mason in Prince Edward Island.[62]

Jarvis died unexpectedly in middle age and left no autobiographical reminiscences like his colleague Hanington. He was a prominent enough citizen that scattered references can be found to piece together a simple picture of his life. But we are fortunate that he kept diaries and that some of them survived and were found, to help flesh out the character of an outstanding, successful, unassuming and memorable person.

In his comparatively short life, Jarvis achieved much and made lasting contributions. He had a remarkably varied career – railway explorations in his 20s, business activities in his 30s and military and police engagements in his 40s. He designed major bridges, helped the initial development of a new university, was involved in municipal politics and was an active citizen in the communities in which he lived. He participated in seminal events in the early history of Canada – the Intercolonial and Canadian Pacific railways, the Northwest Rebellion, the trial of Louis Riel, the boomtowns of Winnipeg and Calgary and the North West Mounted Police. Perhaps he is best summed up by his assistant on the Smoky River Pass expedition, Major Charles Francis Hanington: "I shall still remember Jarvis as a man who knew his business, and who has not been surpassed by any others who followed later on. Jarvis has never sent anyone to do what was his to do. He did it himself, if humanly possible. A most interesting man, who did good work in his profession, and never boasted or employed agents to write him up and keep his name before the public. Peace be to his ashes."[63]

Major Charles Francis Hanington

Charles Francis (Frank) Hanington (see Figure 42) was a member of a Shediac, New Brunswick, family that was prominent in business and politics.[64] He lived through a pivotal period in Canadian history and contributed especially to the development of the national railway network. A year after the Smoky River Pass expedition, in which he was Jarvis's assistant, he described the adventure in a series of letters to his brother Edward. These were eventually deposited at the National Archives and were published in the archivist's *Annual Report* for 1887, together with a map of the route, based on Hanington's original 1875 map.[65] Towards the end of his life, in 1928–1929, he wrote a series of reminiscences for the BC Archives, in which he related events in his long career.[66] The reminiscences have never been published and we know of only one published reference to them, in Pierre Berton's *The National Dream*.[67] Additionally, descendants of the Hanington family have collected many records about their family's history.[68]

Hanington was born on April 14, 1848, the tenth of 13 children born to Daniel Hanington and Margaret Anne Peters. His grandfather William was one of the original English settlers in the Shediac area, acquiring a large area of land at Shediac Cape in 1785 and becoming a very successful businessman. His father, Daniel, entered politics and was the representative for Westmorland County in the New Brunswick Legislative Assembly from 1833 to 1862. Hanington's brother Daniel also went into politics and was briefly premier of New Brunswick in 1882–1883.[69]

Frank Hanington chose a career as a civil engineer. After growing up in the family home, he attended Mount Allison University in Sackville and graduated in May 1871.[70] Hanington immediately obtained a position as a rodman with the Canadian Pacific Railway Survey and was instructed to meet Party M, under the direction of E.W. Jarvis, at Prince Arthur's Landing (now Thunder Bay, Ontario). He left Shediac in mid-June 1871.[71] In his reminiscences he wrote, "Everything was a novelty for me then. I had never seen a lake, and missed the smell of the salt water, and many other things that belonged to the old home life. Seemed like a lost soul, and probably looked the part." For the next two years, he surveyed with Jarvis in northwestern Ontario, summer and winter, with some office work in Winnipeg in the winter of 1872–1873. In June 1873, they were both posted to British Columbia and worked in the Lytton and Lillooet areas. After a winter in Ottawa, plotting their season's surveys, they returned to

British Columbia in 1874, working along the North Thompson Valley, in the Cariboo Mountains and in Yellowhead Pass.[72] It was at the end of this season that Sandford Fleming chose them for the hazardous expedition across Smoky River Pass.

Hanington chose to continue working with the CPR Survey and returned to British Columbia in mid-1875, to survey in the upper Fraser Valley and Yellowhead Pass. While in winter camp at Tête Jaune Cache, in 1875–1876, he wrote a series of letters based on his field notes to his brother, Reverend Edward A.W. Hanington in Ottawa, describing the previous winter's expedition with Jarvis. The letters were published in 1887 as his "Journal." The next two summers were spent in British Columbia, 1876 in the upper Fraser Valley and 1877 in the Fraser Canyon, with the winter in Ottawa.[73] After his return from the 1877 field season, he married Emma Wharton Featherston in Ottawa on November 1.[74] Summer 1878 found him back in northwestern Ontario, revising a line between Kenora and Thunder Bay, followed in 1879 by a shift into railway construction in the Kenora area.[75]

In March 1880, Hanington moved back to British Columbia, where he was employed as an engineer for the CPR construction in the Lytton area, one of the most challenging sections of the route, through the precipitous Fraser and Thompson river canyons (see Figure 43). He and Emma lived in Lytton, and their daughter Emma Wharton Secretan was born in 1881 (in Ottawa) and their son Frank Featherston in Lytton in 1884.[76] Of this period, Hanington said, "I believe the four years I spent on that construction work were really the happiest years of my life."[77]

With the construction work completed, Hanington headed east, first to CPR headquarters in Ottawa, until the end of his contract.[78] Then it was on to Shediac in September 1884, to operate a sawmill that he had bought that had belonged to his brother William who had died in 1883.[79] Unfortunately, the mill burned down in 1885 and was not rebuilt, and Hanington recorded, "I lost $15,000 and that was $4,200 more than I had."[80] He had made a commitment to his parents to remain in Shediac while they were alive and so he took up land surveying.[81] In 1889, he was working as the New Brunswick manager of the De Bertram Railway Syndicate, which was promoting a railway line through the province, connecting to a ferry to Prince Edward Island.[82] His mother died in 1887 and his father in 1889,[83] but Hanington continued working in Shediac, and he was appointed as a justice of the peace for Westmorland County in March 1895.[84] However, in 1896, he went west again and spent a year in the Kootenay district of southeast British Columbia, supervising construction of the Crowsnest Pass Railway in the Moyie area west of Cranbrook.[85]

The next few years were an unsettled time for Hanington. He returned to land surveying in the Shediac area, helped relocate sections of improperly built line for the Bouctouche and Moncton Railway and the Central Railway, both in New Brunswick, and did some surveying in the North Bay area of Ontario.[86] He also returned to the Kootenays (see Figure 44), and in 1899 was superintendent of a mine near Fort Steele, belonging to a Montreal syndicate, as well as the Dupont mine.[87] The whole Hanington family was still in the Kootenays in 1901.[88]

Next, he was hired by the Clergue Company of Sault Ste. Marie (a large conglomerate that included Algoma Steel) to investigate the potential for a railway route via the Moose River to tidewater at James Bay. He recommended against the project.[89] During this work, a narrow escape from death by drowning, in which he went through the ice of one of the northern lakes, crippled his left arm for some months afterwards.[90]

He spent a year on minor railway work at Sudbury, Ontario, then joined the Grand Trunk Pacific Railway, organizing survey work in the Nipigon area but found the work unsatisfactory.[91] He transferred to the rival Canadian Northern Railway (later Canadian National) in 1909, as division engineer based in Vancouver and responsible for location surveys in his old stomping grounds, the North Thompson River valley and Yellowhead Pass.[92] This time, his wife, Emma, stayed in Ottawa, possibly to care for her aging parents.[93]

His next assignment with the company was as division engineer responsible for construction through the challenging Thompson River Canyon in British Columbia, between Spence's Bridge and Ashcroft (see Figures 45 and 46). In Hanington's words: "The construction work was very interesting, and I had some clever resident engineers, a good horse, and of course we were well provided with food and axemen, cooks and a good climate which was hot in summer but never too cold in winter."[94]

The First World War broke out before Hanington had finished his contract, but as soon as he could, he returned to Ottawa, keen to join the cause and do his part in defence of the British Empire, "the greatest empire on earth, and under the most glorious flag that ever flew, for honesty, uprightness, and everything that is good in every way."[95] The fact that he was now 67 years old was not a consideration; he was determined to serve his country.

He obtained a position with the Canadian Hydrographic Service in July 1915, to provide an income while he took training to prepare himself for military service.[96] He was commissioned as a major in the 257th Construction Corps, later the 7th Battalion of the Canadian Railway Corps, an ideal fit for his many years of experience as a railway construction engineer.[97] During this time, he was instrumental in recruiting 250 men for service in his native province of New Brunswick.[98] He was finally able to enlist on January 5, 1917, and gave his date of birth as 1860, thereby shaving 12 years off his true age of 68 to avoid rejection (see Figure 47).[99]

The company travelled to England, then on to Poperinge in Belgium, close to Ypres, although Hanington was not directly involved in any action. His age finally caught up to him: "I was too old when I volunteered to go to the front, and my nerves were pretty well used up when I returned to Canada in the fall of 1917."[100]

On June 10, 1917, Hanington fell ill from exhaustion. He recuperated in hospital until June 17, 1917, when he returned to active duty. Unfortunately, his records had now been discovered as incorrect. On August 26, 1917, Hanington was transferred back to Purfleet, England, as the result of a medical review board's findings. The board had reclassified him as ineligible for active duty due to his age. At the time of this review he was in his 70th year. Hanington finally returned home to Canada on December 4, 1917.[101] After his recovery, he travelled about for a few months, recruiting for the forces.[102]

Even then, despite his age and his wartime illness, he returned to work at the Hydrographic Service in late 1918 and resigned only when forced to, on November 1, 1924.[103] He was 76 years old. He joined his son in California, and after several months there, returned home.[104]

Emma and Frank (see Figure 48) continued to live at 452 Rideau Street in Ottawa, the home that had originally belonged to Emma's parents.[105] He wrote his reminiscences for the BC Archives in 1928–1929, in his own original style. Hanington died on December 21, 1930, and Emma six months later, on June 15, 1931. They are both buried at Beechwood Cemetery in Ottawa.[106]

Hanington's contribution to the exploration of the Rockies is commemorated by Mount Hanington at the east end of Jarvis Pass (Smoky River Pass) in British Columbia (see Figure 49). The mountain was officially named on September 1, 1925, so that he had the pleasure of seeing his exploits permanently honoured during his lifetime. Adjacent are Hanington Pass and Hanington Creek. All are in Kakwa Provincial Park.[107] His contribution is also commemorated by Hanington Road in Hinton, Alberta, near where the expedition reached Jasper House.[108] He is recognized by the Association of BC Land Surveyors as one of the pioneer surveyors in the province.[109]

Hanington was a hard-working and positive person and highly regarded – over the years, various railway companies sought out his expertise. His reminiscences make it clear that he loved his early life on the survey; it was all a great adventure and he lived life to the full. Even after 50 years, he had wonderful memories and wrote of his first season of surveying in British Columbia: "We had a first class outfit, and good health, lots of energy, and good spirits. So, were we downhearted? We were NOT" and "best of all, we always looked ahead, instead of backward, liking to think of HOPE rather than REGRET." He and his party worked together for several years, under all conditions on the trail, summer and winter, until June 1875, when Jarvis left the survey and Hanington was posted back to British Columbia and to another party: "I was quite alone this time, having lost my friends and companions of four years, with whom I had worked and enjoyed it, as well as the pleasant times in which we were in so-called civilization between surveys and explorations. I am referring to Jarvis the Chief, C.E. Perry, J.H. Gray, men who were more like brothers than only friends, each of us knowing his own duties, and doing them without any friction or argument." He and Jarvis loved parties and social gatherings and in his reminiscences he mentions his love of music. Jarvis's diary refers to a party they attended in Ottawa on January 20, 1873, and suggests they were the instigators of four dances in Winnipeg in a two-week period in May 1873, followed by a farewell dance on June 4 that lasted until 3 a.m., just prior to their posting to British Columbia. After their departure, it "felt pretty quiet after the fun of the past two months."[110]

Hanington had a wonderful dry sense of humour that must have been a great asset under trying field conditions. On one occasion, when surveying along the upper Fraser River, he surprised, and was chased by, a bear back towards camp: "I have often read and heard since then that no bear, grizzly, cinnamon, brown or black would ever attack a man, so perhaps my friend Mrs Bear was simply lonesome, and wanted a playmate. I was afterwards told by Mr Marcus Smith that he had heard I took fifteen feet at a stride when that bear was after me, but I never knew where he got that information." He and Jarvis surveyed Yellowhead Pass in 1874 and "Jarvis made the elevation 3,750 feet, and in spite of the fact that very heavy men, in the shape of engineers, presidents, contractors and even cabinet ministers and other very heavy politicians have crossed the divide since 1874, yet I believe the elevation is still about 3,740 feet."[111]

Hanington worked with many Aboriginal and Métis people during his survey work and had a mostly positive attitude, which was not typical of the period. In his first year, he recalls, "there was a guard put on the stores....[T]here was of course no real necessity for a guard at all, but we did not know that the Ojibway Indians had not been Christianized, and were therefore still honest." "On this trip, my trusty rodman and I had our meals and slept at the houses of half breeds along the Winnipeg River, and were well fed and well treated in every way." During his later work with the Canadian Northern Railway, he wrote: "After we finished all I had to do, and got to the next cache, we found that instead of flour, bacon, beans, sugar etc., there was a fifty pound sack of flour and a little tea. The place had been looted, by prospectors of course. No Indian would do such a thing unless he had been Christianized. I have known starving Indians open a cache, take what they needed, put the covers on, and sometimes put in furs to cover the cost of the food they had taken. Not so white men, who leave the caches open to the weather, and ruin the stuff they leave, if any."[112]

Hanington was from a staunchly Conservative political family and his reminiscences reflect the fact. In those days, a change of government could affect the employment of men even at his level, working as lowly surveyors. With the election of a Liberal government in 1878, "the axe fell, my head was off, but I had enough influence personally to get my name reinstated, with a salary of $60 a month instead of $100" for the same work. Later, in 1915, he used his connections to obtain his position in Ottawa. During the Smoky River Pass expedition, when the party ran out of food, Hanington lost a lot of weight. Afterwards, when the government paid them a small bonus, "the Liberal government gave me thirty six dollars for the thirty three pounds of flesh lost, and I considered one dollar nine cents a pound for tough, stringy meat such as mine must have been a blank blank good price for even a Liberal government to pay."[113]

Hanington and Jarvis became good friends, as well as colleagues, during the years they worked together, a tribute to both of them as they were together in the field, 24 hours a day for months on end. They kept in touch after Jarvis left the survey and Hanington spoke admiringly of Jarvis in his reminiscences (see Jarvis's biography). Jarvis, in his turn, in an understatement typical of a formal government report, acknowledged Hanington's contribution to the expedition: "and I must give a word of praise to the pluck and endurance of my assistant, C.F. Hanington."[114] In his journal dated May 22, 1875 (one day after their arrival in Winnipeg), Hanington wrote, "My eye has just caught this sentence in Jarvis' report, which I have been reading (his report to the Chief Engineer, Mr. Fleming): I cannot refrain from mentioning in terms of the highest praise, my assistant, Mr. Hanington, to whose pluck and endurance the success of the exploration is so largely due, I put this in because I am proud of it, and I will add that that one sentence from Jarvis is pay enough for all I did through the winter."

Other Members of the Expedition

Jarvis and Hanington engaged a number of Aboriginal assistants. Apart from Alec MacDonald, expedition members are referred to only by a single name.

Alec MacDonald (sometimes spelled "McDonald") was, in Hanington's terms, "a young Red River quarter-breed, who has been in British Columbia some two years."[115] He was engaged by the party in Quesnel to take care of the dog trains, and was a vital part of it all the way to Winnipeg. Alec led the team that went back from Fort George towards Quesnel to retrieve the caches that had been left en route, a challenging trip that culminated in Jarvis's description of the frozen spectral apparition of Alec appearing at his door. He fell through the ice on Herrick Creek, and was saved by Johnny. He shot a rabbit when starvation loomed for the party, allowing for slight respite, and alerted the party to traces of human activity when on the overland march from the Kakwa to the Athabasca. It was Alec who thought that a peak they saw to the south was Roche Miette, something that gave the starving group fresh hope. He had been through the Yellowhead Pass before, and so was able to be of some assistance in route finding on this section. He assisted Hanington through a state of collapse as they were approaching Lake St. Anne. After the expedition, Hanington noted that Alec was hired as a mail carrier between Winnipeg and Edmonton till the autumn, after which he was sent to Henry House (Jasper) to look after the depot there. He had an enjoyable visit with Hanington over the winter and they reminisced and celebrated the anniversary of the day they had reached Lake St. Anne. The plan then was for him to return to Winnipeg the following spring.

There is one further enigmatic reference to Alec. On August 11, 1875, Jarvis was in Winnipeg, and noted in his diary as he arranged for Alec to return west: "Arranged with Rowan to send Alec McD to take up cache on Smoking River." Whether or not he was able to perform this duty is not known.

Like Alec MacDonald, Johnny was engaged by the party in Quesnel and served with it all the way to Winnipeg. Part of his role was cooking. On one occasion, he saved Alec through quick thinking by pulling him out when he fell into Herrick Creek. Hanington approvingly described him as "unmoved as always," "as good-natured as ever," and "silent and impassive,"[116] even in the face of great distress. Hanington related that Johnny's constant sentence was "*Cultus kopajnika. Cultus kopa mika*": "What's bad for me is bad for you." Hanington provides a description of a decked-out Johnny involved in a courtship in Winnipeg after the trip, his board paid for. He visited Hanington late that fall on his way back to Quesnel, and told him he would hopefully soon be a father.

The party engaged Quaw in December 1874 in Fort George. He helped retrieve some of the caches left between Quesnel and Fort George. He had a cache of fish on the Bear (Bowron) River, and it was planned to purchase this from him to supplement the food supply. He proceeded with the party from Fort George to Hanington's Cache, assisting en route in the shooting of a grouse. He then led Hanington and Te Jon to the salmon cache, where he had a cabin. He provided Hanington with verbal directions on the way through the mountains (including passing a "falls as high as a tree"), which were to prove problematical in the days ahead. Hanington and Te Jon said farewell to Quaw at this cache. Members of the Quaw family still live in the Prince George area.

The party engaged Te Jon, whom Hanington referred to as an "Indian boy," in Fort George. He joined Hanington and Quaw on the side excursion to the salmon cache on the Bear (Bowron) River. He assisted Hanington on the return trip to link up with the main party, and participated in the expedition up Herrick Creek. From the forks of the McGregor River and Herrick Creek, he and Tsayass were sent back to Prince George with one sled and seven dogs, picking up extra provisions en route at Hanington's Cache. "Te Jon" is the colloquial pronunciation of Tête Jaune, the small settlement at the western entrance to Yellowhead Pass. The name originated with the famed Iroquois hunter, trapper and guide Tête Jaune (Yellowhead), who was active in the region in the early 1800s. It is possible that Te Jon was a descendant.

Hassiack (Assiak) and Ah kho (Ahiko) joined the party at Hanington's Cache, having come with Tsayass from "Stewart's Lake" (Fort St. James) with three dog trains and supplies. With one exception (Ah kho going with Alec and Johnny up the McGregor River to make a trail), they are referred to in both Jarvis's report and Hanington's journal simply as "the Indians." They formed part of the expedition up and back down Herrick Creek, then all the way up the McGregor River, through Jarvis Pass, down the Kakwa River, overland to the Athabasca River and then east to Fort Edmonton, where they were left in the care of the chief factor, with plans for them to return to Fort St. James in the spring via Yellowhead Pass. Jarvis wrote: "My Indians at times became much disheartened, but behaved well throughout."[117] Hanington was critical of their demeanour when their fate was uncertain and they believed their end was near.

Tsayass, with Ah kho and Hassiack, drove three dog trains, stocked with supplies, from Stewart's Lake (Fort St. James) to Hanington's Cache, where they met the party. He partook in the venture up Herrick Creek then was sent back to Fort George with Te Jon.

John (Jack) Norris (1826–1916) joined the expedition in Fort Edmonton, and helped guide the party all the way to Fort Garry. Born in Scotland, he joined the Hudson's Bay Company and worked as a labourer and boatman. He had helped lead the first brigade of Red River ox carts from Winnipeg to Edmonton in the early 1860s, a journey that took three and a half months. Thus, he became an experienced guide.[118]

Sandford Fleming

Sandford Fleming was not a member of the expedition, but the influence of "The Chief" is felt throughout the writings of Jarvis and Hanington – their sense of obligation and duty, their loyalty and respect. Fleming was born in Scotland in 1827.[119] In his very first diary, as a young boy, he copied a quote from *Poor Richard's Almanack*: "But dost thou love life? Then do not squander time for that is the stuff life is made of."[120] Fleming put this adage into practise. By the age of 18, he was a qualified civil engineer and surveyor. The opportunities and excitement that Canada offered beckoned, and in 1845 he set sail for Peterborough, Ontario. From 1853 onwards, railways dominated his life. He climbed the ladder to become chief engineer of Ontario's Northern Railway from Toronto to Collingwood. When few sane people seemed to be contemplating it, he was already thinking about, talking about and making detailed proposals on a railway right across Canada to the Pacific; in fact, he produced a treatise on this topic, calculated to the last dollar, at age 35.[121]

In 1867, he was appointed chief engineer of the proposed Intercolonial Railway from Quebec to Halifax. The project was ambitious and important: creating the critical physical link between Upper and Lower Canada and the Maritime provinces. In fact, it was a condition of Confederation. As chief engineer, he then got the railway built, and in the summer of 1876 was able to declare this megaproject complete. That was more than a year after Jarvis and Hanington completed their epic trip.

It strains credulity, but in 1871, with his hands already full with the Intercolonial Railway, he was appointed engineer-in-chief of the proposed Canadian Pacific Railway, the item that, together with the Intercolonial, would bind the Dominion of Canada from sea to sea. Fleming seemed to have realized the virtual impossibility of attending fruitfully to both positions, and tried to decline the offer. However, it became more of a command, a reflection of his duty to the giant yet fledgling nation, and he felt obliged to acquiesce.

For five years, then, he grappled with both these arduous, demanding jobs, the one supervising to completion a daunting task, the other contemplating and turning into reality a dream that many considered to be lunacy, and what at the time was without a doubt the most ambitious railway project ever undertaken.

In 1885, Sandford Fleming was with William Van Horne in one of Canada's most celebrated photos, observing Donald Smith driving in the last spike of the Canadian Pacific Railway. By that stage, a lot had happened – Fleming was no longer in charge and his choice of the Yellowhead Pass through the Rockies had been ignored for political reasons.

There was a lot more to Sandford Fleming than railways. He pioneered the idea of time zones, led the drive for the Pacific Transoceanic Cable to unite the British Empire, founded the Royal Canadian Institute, cofounded the Royal Society of Canada, became a captain in the Canadian military, tended to a large extended family, designed Canada's first postage stamp (the Threepenny Beaver, designed in 1851) and published a book of prayers. And, incidentally, he managed to help save the portrait of Queen Victoria that hung above the throne during the burning of the Parliament buildings in Montreal.

He was let go in 1880, with what today would be called a golden handshake. There was some justice to this, as for years he had been paid for doing only one of his two simultaneous jobs.

He received a knighthood from Queen Victoria in 1897.

APPENDIX A

The Smoky River Cache

After crossing Smoky River Pass and deciding to continue to the Athabasca valley and then on to Edmonton, the expedition members travelled down what is now known as the Kakwa River, a tributary of the Smoky River, for approximately 140 km, until they were certain that it was heading northeast, away from their destination. It was a long detour. In his narrative report, Jarvis wrote:

> [W]e knew by our latitude, obtained by observation, and our approximate longitude, calculated by dead reckoning from the track survey we were making, that our course to strike the Athabasca River and the country we wished to explore between it and the Saskatchewan River would be about south-east, while we were now travelling at right angles to this course, or north-east. I could not, however, abandon the hope of the river shortly turning to the east, or even more in the desired direction, so we held on a few days longer.... A long and earnest consultation, in which three different propositions were made: 1st, to assume we are on Smoky River, and to follow it to Peace River and Fort Dunvegan; 2nd, to go east to Fort Assiniboine, on the Athabasca; and 3rd, to go south-east to Jasper House, ended in the adoption of the latter; and the following day, finding the river turned still more to the north, orders were given to camp early, and a suitable place chosen to build a cache in which to leave everything that could possibly be spared.[1]

By this time, they were low on food and had lost many of the dogs. The remaining dogs were too exhausted and too few to pull the sleds. Jarvis decided to jettison everything even remotely nonessential. He reported that "the other two sleds and their harness, together with superfluous clothing and instruments, were placed in a small log hut, six feet by four, and three feet high, built for the purpose, and the names and date marked on surrounding trees."[2]

In striking southeast, they were committing themselves to harsh overland travel, against the grain of the land. Following a frozen river downstream may have seemed simpler and quicker.

Why, then, the decision to head southeast? What would have happened if they had simply continued downstream? The Kakwa River joins the Smoky River, and their combined waters join the Peace River near the present-day community of this name. Upstream on the Peace River was Fort Dunvegan, established by the Northwest Company in 1805 and still the major outpost in the region.

But in 1875, heading east from here to their ultimate destination of Fort Garry (Winnipeg) would have been challenging. The fur-trade route with its portages may have been suitable for voyageurs in summer but would have seemed almost insurmountable to the hungry, nearly exhausted party. A more feasible option would have been an overland route from Fort Dunvegan directly east to the Hudson's Bay post on the shores of Lesser Slave Lake, then east again on a difficult route to Fort Edmonton. Compared with either of these options, the nearer Jasper House, with its promise of provisions, must have seemed more attractive.

Besides, there was a pressing reason for heading east, consistent with their loyalty to Sandford Fleming and their surveying duties. They had been instructed not just to search for a pass through the mountains but to then explore in an easterly direction. In his report, Jarvis would write: "I was unable to explore a line from Root River to White Earth (old) Fort, as you directed, owing to extreme bodily exhaustion consequent upon the hardships we underwent."[3] Duty called.

With his usual attention to detail, Jarvis left a description of the contents of the cache in the back of his diary:

6 yds tartan (Ahiko)
1 carpet sack (Ahiko)
2 plugs tobacco (Ahiko)

1 gimlet (Ahiko)
2 prs moccasins (Assiak)
1 merino shirt (Johnny)
1 skin pants (Johnny)
1 old towel (Johnny)
2 old shirts (Johnny)
(12 lbs)

1 sextant (containing artificial horizon) 15 lbs
1 stationery box containing:
- Medicine
- Pencils
- Bill forms & receipt book
- 1 4-inch compass, folding sights
1 boiling point thermometer, complete
1 pocket sextant
1 field glass
1 min. thermometer
1 maximal thermometer
1 50 ft tape
1 pair scissors
1 handbook (C.F.H.)
1 "medical guide"
1 file
2 awls
Red chalk
(35 lbs)

1 bag containing ictas ["things" in the Chinook language] and
6 pairs moccasins (10 lbs)
Rifle & pistol cartridges (12 lbs)
1 tin box, containing
- 4 boxes yeast powder
- 2 lb. tea
- 3 lb. coffee
(7 lbs)[4]

The report, the diaries and Hanington's map together help identify an approximate location for the cache. Jarvis calculated that they travelled down the Kakwa River for 87 miles (139 km) from Smoky River Pass. Even allowing for their following every bend in the river, this is questionable. A comparison of his listed distances with modern maps suggests they overestimated by up to 25 per cent using their simple counting method. A corrected distance would be closer to 65 miles (104 km). Jarvis left a clue in his diary: "Followed down the Smoky R. a mile, and turned off to the S. up a 30 ft. creek."[5] (In his report, he claimed this was a few miles, but the diary entry is likely more accurate). Using all the available clues, it is possible to make a best estimate of the cache location: 54° 15'N, 118° 55'W. This location is close to the confluence with Prairie Creek, in the lower Kakwa valley, a short distance north of Grande Cache.

One of the unknowns is the fate of the cache. On August 11, 1875, Jarvis, arranging to dispatch Alec MacDonald back west from Winnipeg, wrote in his diary: "Arranged with Rowan to send Alec McD to take up cache on Smoking River."[6] Whether or not this happened is not known.

There are certainly enough metal items in the cache, as detailed by Jarvis, to justify using a metal detector in a search for this fascinating item of Canadiana.

APPENDIX B

The Jarvis Diaries

It is not known how consistently Jarvis kept a diary, but we are fortunate that seven have survived from different periods of his life. They have been particularly useful in preparing his biography.

There are two interesting stories associated with the diaries – their journey from Jarvis's possession to the Archives of Manitoba, and the trail that led to their rediscovery for this book. And, intriguingly, there is also a link with the long-lost diary of Louis Riel.

Jarvis presumably kept his own personal diaries throughout his life. The Riel diary, as related in his biography in Chapter 6, came into his possession immediately after the Battle of Batoche and for some reason was not forwarded to Ottawa, even though it was used as evidence in the trial of Louis Riel. At some point after 1890 (the date of the last diary), all the diaries ended up at the Alloway and Champion Bank in Winnipeg, possibly in a safety deposit box or in the personal care of the two bankers, who were Jarvis's executors and presumably personal acquaintances from his days in Winnipeg. Following Jarvis's death, Henry Champion wound up his estate but did not forward the diaries to his family. Years later, when the bank was moving to new premises and discarding unwanted items, a young accountant, George Waight, was permitted to keep an old, rosewood, roll-top desk from the basement. And when he eventually started to restore the desk he was astounded to discover the diaries behind one of the drawers. Fortunately, he recognized their significance and ensured that they went into safekeeping. The Riel diary is now preserved at the

University of Saskatchewan Library and the Jarvis diaries are kept at the Archives of Manitoba.

When we started researching the story of the expedition, we were unaware of Jarvis's diaries. We owe a big thank you to Lena Goon of the Archives of the Canadian Rockies in Banff, who drew our attention to *The Young Surveyor*, a young adult novel about the expedition by Winnipeg author Olive Knox. She dedicated the book "To Ellen and George Waight: In appreciation of the use of Edward Warrel [*sic*] Jarvis' original diaries and Field Books, from which this story was created."[1] That was the eureka moment – there were original diaries – somewhere. But where? Knox's book was published in 1956, more than 50 years previously, so it was unlikely that she or the Waights were still alive. An Internet search of online databases of archives across Canada turned up nothing. It was the same story with the University of Manitoba Special Collections and the Manitoba Historical Society. So it was back to the Internet to try to find the Waight family, in case the Jarvis diaries were still in their possession. The search was a sobering lesson about the lack of anonymity in the digital age. We found an obituary for Ellen Waight that gave us their daughter's married name. And then we found an obituary for their son-in-law in which his wife's maiden name provided the link back to the Waights. Once we had confirmation of the Waights' daughter's married name, we were able to trace her via a newspaper article to a seniors' residence in Winnipeg, and from there to her daughter. They told us the story of how the diaries were found and confirmed that George Waight had transferred them to the Archives of Manitoba. So it was back to the archives, this time with a direct inquiry, and we received the information that the diaries are indeed kept there but are not included in the online catalogue.

The search for the Hanington reminiscences was easier. Our first knowledge of their existence was a brief reference in Pierre Berton's *The National Dream*, which led us to the BC Archives. Once again, we discovered an example of material that is in a collection but not listed in the online catalogue.

With the assistance of the two archives, we have been able to highlight the existence of the diaries and the reminiscences and publish excerpts for the first time.

APPENDIX C

"Winter Journey"

BY JON WHYTE

This poem by Jon Whyte was published in *Tales from the Canadian Rockies*, edited by Brian Patton and published in Edmonton by Hurtig Publishers in 1984. Jon Whyte died in 1992. The poem is published here, with the permission of Jon's brother, Harold Whyte.

1)

January sets in very cold.
We redouble our exertions to prepare everything.
Thermometer down among the forties,
one six a.m. marked fifty-three below.

2) January 8th

Alec appearing from the darkness, overdue,
moving slowly, weary, livid, drawn,
face, flesh, clothing, hair, parka trim
a frosty bristling, a silvery pallor;
winter in every pace
in his near-dead trudging.
Seeing him, ourselves foresaw;
the season getting on, we must be on our way.

3) January 14th

The sled dogs fossick warmth,
curling into each other and the snow.

No wind;
all still.

4)

The largest, most roaring fire
little more than burns the side toward it;
the other, spitted to darkness, freezes.

5)

Were it anything except this cold, this glimmering,
we might retreat to sleep to dream of warmth,
mutter, "This is the maddest madness," and turn back.
Obliged to thrust on,
to measure how far, how near, how deep, how clear
Death's soft step sounds,
we won't turn back.

6)
January
15th

In cold's depth
the fire steams and spits; it does not smoke,
the driest wood we find.

Frost quills bare flesh.

7)

Alec suddenly and silently appears,
epiphany in frost,
knew his route, where he was going, fell into streams.

We cannot tell in cold precision whither we go.

8)

Days so short we rise in darkness to revive the fire,
melt snow for tea;
while we drink it, fold our blankets,
take down the fly, harness the dogs;
in dawn's deep indigo strap snowshoes on,
move out to meet it.

9)
January
16th

"I can't imagine a quicker way to harden a man's heart
than to put him driving dogs."

Expected not so early on our way,
Death seized old Marquis
who limped all morning our third day,
at noontime wagged, then rolled to darkness,

Jarvis reports, the official version,
noting the day forty-six below,
echoing involvement and discretion.
They buried Marquis in a grave of snow.
Marquis – leader – then Cabree, Sam, Buster,
dogs lively "as crickets" *at feeding time*,
writes Hanington. Marquis lacklustre,
iced, frozen to shoulders, thick in rime,

had to be shot;
Jarvis did the deed.
Whichever: we left the good old brute
more at ease than since he froze his feet.

10)
January
20th

The snow lies deeper as we near the mountains,
turn up the Fraser's north branch
to where we hope our Smoky River pass shall be.

11)

The fellows in Quesnelle expressed
exalted notions of the pleasures we'd derive
from our winter journey through the mountains:
"God bless you old fellows – goodbye," they said.
"This is the last time we will see you."

12)
January
29th

We reload, relash sleds, four in number,
twenty-four dogs, eight of us,
mend moccasins and harness, glum in our
prospect of retreat; forward we must.

In blue and silver dusky light
hearing distinctly a chopping axe
on the far shore as we turn in, the bite
of blade, the wrench away: a fact

none will leave the fire or into cold dash
to solve: the mystery of our axeman company.
It deepens when we hear a large tree crash
branch-shatteringly. We sit, accepting and denying.

Morning, we see no fallen tree,
no notching blaze, no fallen branch, no print
explaining the sullen mystery.
Imaginations, spurred by nothing, will not rest.

13) Hard work, breaking track;
anything to think of is pleasanter;
walking all day, thinking of nothing but "1, 2, 3, …"
is monotonous for anything.

14) Tiger: shot, the 29th.
His lameness prevented his doing anything
except eat grub.

15) February 4th: a heavy snowstorm
just to make things lively.

16) February 8th
Provisions for another month, no more;
no notion of how far we have to go.

One goes ahead setting the track,
flailing down the snow,
creating a trough
where the dogs can pull our sleds.
Where the canyon walls lean in on us
we abandon the stream to portage,
clamber up the walls
haul dogs and sleds too,
two men hauling, two men pushing.
Where the descent begins, returning to the stream,
the sled invariably tips,
pulling the dogs down with it.
Poor beggars: we cannot rest them or ourselves.
Each step an agony, snow sucking under,
caving beneath our feet which crave solidity;
we haul one up, the other steadying itself
on soft, collapsing crust.
One by one the river's branches we explore,
preferring none.

Mountains conceal their passages and passes
behind the steepest walls, the deepest canyons,
the most impenetrable portages, the thickest hollows
where snow piles up in deep soft sift
the wind has borne down watercourse.

We lack days to waste traversing an unknown,
apparently unlimited distance on a known
shrinking store of dry-flake salmon.

Plunging into the snow, our feet each step
create small avalanches from the toe.
Our feet are blistering;
nausea swims in our heads.

A mile more: walls of rock we cannot climb
confine the river's and our course.

We seize on any coign of vantage,
narrow ledges, banks of ice,
frost bridges from one boulder to the next,
black water boiling, foaming at our feet,
about to snap up, swallow any who slip.

Canyon succeeds canyon;
river bed so full of boulders,
our progress treacle slow.

The weather stormy,
snowfalls frequent, snowshoeing laborious,
our spirits sink.

In camp we shovel down to moss,
the snow so deep
we cannot peer from out the pit we dig.

17) In stillness is beauty.
In beauty is death.

18) The blades of the mountains sharpen distinctly
in the bright peach-golden light dancing to day,
the sky pale blue;
when shadows stretch to night: a spangled place.

Great glittering blue
glaciers ride the ridgewalls far above us.
Jarvis thinks it would be beautiful
to see it from the comfort of a Pullman.
Each stream branch comes to this:
an amphitheatre of high bare rocky peaks
where slender clear-blue lines of glaciers menace.

Hanington says:
"We seem to have got to the back of the north wind."

19) February 13th
How many ways, how many branches,
how many dead ends can streams flow from?

Quaw said three days' travel, a fork to the left,
two days more, a fall high as a tree
we'd portage around;
five days, to a meadow,
three days' travel more,
a stream runs east, where we will see
the sun rise from the prairie.

Quaw's prediction: thirteen days, the height of land.
Eighteen days' travel back.

20) In canyon's depth the gorge engulfs both sky and earth.
The rock leans in so doubtfully
Hanington creeps on it hands and knees toward a fall
"high-all-the-same one stick."
The snow collapses;
limpet-like he clings to rock
until we find a pole to support him.

21) Retreating to a pyramid portage, ascending,

we haul the dogs, whip, denounce, cajole,
reach its summit, then descend and
cannot stop and in a melee roll.

A good man could disentangle them
without a whip and swearing.
We cannot be that sort of man,
nor keep a nerveless bearing.

At one hill's top the dogs surge up and disappear,
falling pellmell in their downward career,
till at a meagre sapling the sled shears
to one side, the dogs to the other veer,

dangling by traces in ludicrous plight,
swaying up and down, bobbing in the air,
like sleigh and reindeer caught in flight
by a chimney on a Christmas night.

22)

The day before the canyon
we sent back to Fort George
two Indians with seven dogs, one sled.
Sam and Chun escaped their keepers,
came back to us.

Our salmon running low,
Jarvis shoots them both.

23)
February
20th

To bed in all available clothing,
the thermometer at forty-two below,
we wake to rain pattering our faces;
within eight hours.

24)

Branches of possibility extend like arms of aspen:
from trunk to stems, from stems to branches,
branches to yet smaller branches,
filling out the tree of which one twig is one,
the branch seen from within.

The branching nerves of bodies form a tree;
from brain to trunk or branch
one touching nerve or point, the pass.

25) "Give it up?"
"You know what I was thinking?"
"For God's sake, yes, we'll not go back."
"If we all come to grief, I am responsible."
"I'll take my own responsibility."

We'd sooner be found in the mountains.

26) February 24th

Grand pyramids of peaks the entrance guard,
glacier most magnificent, transparent blue,
we fancy seeing rocks beneath and through it.

27) Thunder rolls, our eyes peer high:
masses of ice and rock leap
point to point, spout from the sky,
a weird, gigantic fountain heaps

itself, burrows out anew, then falling;
a boulder plunges straight toward us where
we stand, wraith of nature so appalling
we did not make our camp near there.

28) Cold exhilarating morning:
five paternoster lakes, highest in the mist.
joyously we hail a tiny trickling stream,
the sweetest thing, running to the east.

We stop to wash our hands and faces,
the second time in our long journey,
to wash all B.C.'s grime and dust away.

29) Eastward ho! Our spirits rising as we drop.

30)
February
26th

The river bends sharply: an abyss yawns before us:
a fall two hundred feet.
Had the day been misty, we'd have plunged with it.

31)

Scarcely a day passes when their dismal howl
does not announce to our unwilling ears
another dog has dropped.
A trivial incident like the death of a dog
(such mongrel curs as some of ours)
would not affect us in a civilized community.
Here it casts a gloom.
Even the dogs look at one another
as if to say, "It may be my turn next."

32)

Jarvis's face white, lips set.

A reeling, swirling, angering pain
rises in the wrinkling of the brain,
sways with each rough slouch forward
until our muscles knot, spine tears apart
in every jolt, then settles down relentlessly
while nerves their lacerations start anew
cold and numbness will not stanch,
a boring up the vaulted tendons of our feet,
a clawing up through our clenched calves,
whilst all the while our minds rebel,
call out to vomit, retch, heave up, to stop
the swarming and miasmal blurring of our way.
Mal de raquettes: the snowshoe torture.

33)

Knew we which river we proceed along
we could be happier.
It heads east, we think the Athabaska,
our hearts beat high.

It turns north, it is the Smoky,
Our spirits drop to zero.

34) Nothing of the world remains:
warmth went;
familiarity;
the places we might have known;
shelter, time, and banter vanish;
thoughts we might have had;
all but the counting fades: "1, 2, 3, 4, 5…"

35) Dead reckoning,
our course lies south and east;
we proceed by north and east;
holding our hope the river will turn east,
we keep our course a few days more.
It comes to nought.

We turn south.

36) March 6th
A day of rest:
we cache our instruments,
extra clothing we no longer need,
mark our names, the date, upon a tree.
Tomorrow we head south,
depart this valley leading northward and away
from any hope of meeting Indians who've wandered here
in search of game.

37) *Cultus copa nika; cultus copa mika.*
What's bad for me is bad for you.

To stay put too long is death.
To slip and break a leg is death.
In stillness: death.
A silent axeman follows us,
leaves no footprint in the snow.
The silent man is following us.

Hunger burdens our packs.

38) Two hours before sunrise,
our usual time to get going,
when we can see to put one foot before the other.

The frequent ups and downs wear on the dogs;
grown weak, they fall.
To cease their suffering
the stragglers receive the *coup de grace*,
Their bodies heap up on the sleds;
the others closing up, continuing,
at night howl requiems for dead companions.

39) March 11th

Alec shoots a rabbit:
quite a feed for six men.

The Indians in mournful state weep bitterly,
declaring we are lost,
they'll never see their homes again.
Persuasive eloquence is difficult when we're uncertain
our reasoning is sound.

40) Buster, a favourite among our dogs,
we cannot coax to leave the fire.
No one has the heart to shoot him.
We leave him to his fate.

41) Thick mist shrouds everything:
we grope in darkness.

Times are hard when we eat dog to keep our strength up,
dog which has starved, which we have worked to death.
Dog soup may not taste good, but it goes well.

42) March 15th

The rising sun dispels the mist.
The snow has stopped.
Twenty miles away a high bold rock,
much like a photograph once seen: Roche à Miette.
If the Athabaska be not in that valley,

it lies beyond those mountains.
We lack both grub and strength to carry us across.

Alec, hoping we do not see him, steals from camp,
to assure himself by moonlight he's not mistaken.
43) Imagine our camp then:
opposite sit the Indians:
Johnny, silent and impassive,
the other two with their heads in their hands sobbing.
Jarvis, very thin, very white, very much subdued.
Alec chewing tobacco and looking about used up,
not sure if it be Roche a Miette or not.
In the centre I sit.
My looks I can't describe;
my feeling scarcely,
I don't believe the Athabaska lies in that valley.
I do believe we have not many days to live.
I have been thinking of "the dearest spot on earth to me,"
of our Mother and Father,
of all my brothers and sisters and friends,
of the happy days at home,
of all the good deeds I have left undone
and all the bad ones committed.
I wonder if our bones will ever be discovered,
when and by whom;
if our friends will mourn for us long,
or do as is often done,
forget us as soon as possible.
I have been looking death in the face
and have come to the conclusion
C.F. Hanington has been a hard case
and I would like to live a while longer to make up for it.

I am glad since we started
we didn't go back;
this has been a tough trip,
this evening the toughest.

44)
March
16th

Three miles from our camp
the benches of our long-sought river.
The frail dogs stagger, barking feebly;
the effort too much for one, he drops in his traces.

On Lac à Brule, its snow blown off, the ice a glare,
we travel without snowshoes
our first time since Christmas
upstream to Jasper House
where, we hope, we can provision us.
The old fort leans most doubtfully.
It is abandoned, and we've come these twenty miles
for nought.

45)

Some Indians we meet sell us their last provisions:
sixty pounds of dried deer meat.
With one day's rest we start again.

46)
March
24th

From Mcleod Portage
our last view of the Rocky Mountains:
few among us loath to turn their backs on where we
toiled:
bounded by lofty crests, the snowy peaks,
more beautiful by rosy hues of rising sun,
more and more interesting as we depart them,
shaking the snow from off our feet against them.

47)

Tea now is everything;
boiled over and over carefully,
with one fresh grain each time.
Tobacco in the evening improves on nothing.

48)
March
27th

Numbness holds our limbs,
we cannot push one snowshoe before the other,
as if we're marking time.
No laughing matter,
it could amuse a well-fed bystander.

49)
March
30th

Despite our hunger and our weakness,
we stumble into Lake St. Anne.
McGillivray sets a supper out for us,
white fish and potatoes, milk and bread, sugar, tea.
We eat a great deal more than can be good for us.
For half an hour not a word;
then we cannot mumble much.

Five a.m.: we rise and steal some bread.
At seven: breakfast.
Then to a mission village where we buy butter, eggs,
feed on grilled buffalo bones they serve us.
With eggs, butter, cream, we return to the house
where we eat bread with cream and sugar up to noon.
At noon: another fill.
Afternoon and evening through we eat,
as hungry yet as ever.

Our great exploration at an end,
our hunger and great danger.

"It is altogether too large a country for six men."

50)
May
22nd

We had taken the sun's altitude each noon,
determining our latitude.
Using a compass, counting paces for dead reckoning;
sticks in our hands to knock the slush
from off our snowshoes every step, and counting:
at forty paces closing the left hand's little finger
– a hundred feet;
another forty paces, ring finger closing
– another hundred feet.
When we'd clasped closed our left hand's fingers and the
thumb:
– five hundred feet;
shift the stick to the left hand,
start counting on the right.
When we'd clenched that fist,
we tallied on a thousand paces, notebook entry.

Somewhere on the plains,
when we and spring had both run out of snow,
we nailed our snowshoes to a tree,
appropriate inscriptions on them.

We reached Fort Garry, bathed and changed our clothes.
Walking on the hotel verandah:
"How far do you make it?"
My hand was clenched,
my fingers acting automatically,
closing automatically at every forty paces.
Funny, eh?

APPENDIX D

Excerpts from Other Writers

The published accounts of Jarvis and Hanington appeared 11 years apart in obscure government documents, the reports of government staff doing their jobs. Nonetheless, those who knew of their exploits were full of admiration. Sandford Fleming, who had assigned them the challenge, was especially appreciative and later made a point of highlighting their work. We have mentioned the comments of Douglas Brymner and Gerry Andrews.

A contemporary who also wrote of their expedition was J.H.E. Secretan, whom they had briefly met in Quesnel before their departure. He was a friend and supporter of Sandford Fleming, and played an active part as a civil engineer in the Canadian Pacific Survey. He cut his teeth working on railway surveying north of Lake Superior in adverse winter conditions. He became part of the N Division in 1874, surveying from Fort George up the Fraser Valley. He was an able raconteur, and in 1924, his best-known book was published, one that for decades was the most readable account of the "national dream": *Canada's Great Highway: From the First Stake to the Last Spike*. In it, there is a brief reference to the chance meeting with Jarvis and Hanington:

> In order to emphasize these opinions of mine I must not forget to remark that in the fall of 1874, when on my way home, passing through Quesnelle, I shook hands with E.W. Jarvis and C.F. Hanington and wished them good luck. They were fitting out for a winter trip through the Smoky River Pass, which had been reported as feasible.

> They had got together some Indians and about thirteen dogs of different denominations. Although suffering untold hardships they accomplished that long trek across those icy barriers and having eaten their last dog arrived at Edmonton in the Spring of 1875, very emaciated but alive. And I have no doubt the report of these gentlemen eventually found its way into the blue books of Canada, but their personal loyalty and bravery can never be over-estimated.[1]

His words have the ring of authenticity and empathy. He had been on the verge of starvation himself in desperate circumstances, surveying in Ontario in winter. He understood as well as anyone what they had been through.

Another admirer was Charles Frederick Carter, an American railway historian, who in 1908 published the book, *When Railroads Were New.* The first eight chapters deal with early railways in the United States. The book ends with Chapter 9, titled "Romance of a Great Railroad," which features the CPR. It is 29 pages long, and fully a quarter of it is devoted to the Jarvis-Hanington 1874–1875 winter expedition. He effusively summarized their expedition, with colourful passages such as: "It was now a scramble for life over an exceedingly rough country, with the certainty of death staring them in the face if they failed to find, within a short time, the one place where there was help. Every morning one or more of the dogs was too weak to rise and had to be shot to cut his sufferings short."[2]

A fictional account of the expedition has also appeared. As mentioned previously in Appendix B, Winnipeg writer Olive Knox published *The Young Surveyor* in 1956. She was a teacher, a graduate of the University of Manitoba, who travelled extensively and wrote stories, plays and books. This is a young adult novel in which a couple of teen protagonists take Hanington's place as Jarvis's assistants. The novel remains true to the main events as Knox had access to Jarvis's diaries. The back cover of the dust jacket explains:

> What boy or girl wouldn't like the adventure of surveying in the Rockies when pack-mules, gold miners and railway surveyors followed trails known only to Indians? Bruce, the greenhorn from San Francisco, and Josette, the French girl from Edmonton, and her half-brother Mike had that exciting experience when they

> accompanied Edward Jarvis on his Canadian Pacific Railway surveys in 1874–1875. A rock slide, plunge into a treacherous river, forest fire, cougar attack, snow avalanche and losing the trail are some of the dramatic events in this book for boys and girls. This was the life of the first Canadian Pacific surveyors, who pioneered in trail-finding so that the Atlantic could be linked to the Pacific with steel.[3]

At the back of the book, in the "Author's Note," Knox wrote:

> Edward Jarvis later became a Mounted Policeman, travelling the trails of the North-West Territories as he had the mountain trails when an engineer. He died in 1894 but is still remembered as one of the great trail-breakers, and railway-surveyors of Canada...The author hopes that this novel which is, in the main, based on facts, recalls many young engineers, like Edward Warrell Jarvis, who explored new territory, made instrument surveys, and carved the trail over which locomotives now whistle their arrival and call others to come and see this vast and beautiful country. In doing so the readers may remember the dangers risked by the men who mapped out every foot of the road over which they can speed in luxurious comfort. Long live the heroic railway surveyors in our memories![4]

Jarvis is unquestionably the adult hero, even making a dramatic speech and calling for a recitation of the Lord's Prayer at the summit of the pass.

Pierre Berton (1920–2004) was one of Canada's most popular authors of nonfiction and Canadian history. His output was prolific – over 50 works extending over more than half a century. His attention to Jarvis and Hanington featured in his 1970 book *The National Dream: The Great Railway, 1871–1881*. The detail he devoted to their tribulations signifies a sense of awe at their achievements and survival. He knew how to summarize pages of reports and journals into choice phrases that encapsulated the hardships and entranced the reader:

> Both Jarvis and Hanington left graphic accounts of the ordeal, illuminated by uncanny episodes: the spectral figure of Macdonald knocking on the door of their shack in 49 below zero weather, sheathed in ice from head to toe; the lead dog who made a feeble effort to rise, gave one spasmodic wag of his tail and rolled over dead, his legs frozen stiff to his shoulders; and the auditory hallucinations experienced one night by the entire party – the distinct

> but ghostly sound of a tree being felled just two hundred yards away but no sign of snowshoes or axemanship the following morning...They moved through a land that had never been mapped. A good deal of the time they had no idea where they were...They fell through thin ice and had to clamber out, soaked to the skin, their snowshoes still fastened to their feet. One morning, while mushing down a frozen river, they turned a corner and saw an abyss yawning before them: the entire party, dogs and men, were perched on the ledge of a frozen waterfall, two hundred and ten feet high; the projection itself was no more than two feet thick. One evening they made camp below a blue glacier when, without warning, great chunks of it gave way...A chunk of limestone, ten feet thick, scudded past them, tearing a tunnel through the trees before it plunged into the river.[5]

In 2001, the British Columbia Historical Federation commissioned Yvonne Klan to summarize Hanington's journal, which had come to the attention of Barry Cotton of the Association of BC Land Surveyors while researching early surveyors in the province. The article's introduction refers to "a horrendous 1874 exploration headed by E. Jarvis, with Hanington serving as his assistant." Cotton wrote: "The Jarvis-Hanington exploration has been all but forgotten and deserves to be made known."[6] Klan agreed and accepted the task of whittling down the account from 9,000 words to 3,500, although she did not provide any additional information.[7]

Jarvis's own official account of the expedition was summarized by well-known Alberta historian James McGregor in the *Alberta Historical Review* in 1958.[8]

As recently as 2011, the expedition was singled out in *Great Railroad Tunnels of North America* by well-known climber William Lowell Putnam: "[I]t is clear that railroads, and particularly reports on their location and construction, have thus often become primary sources for the study of early North American mountaineering and exploratory ventures. One such venture was the lengthy tour de force by dogsled undertaken by the redoubtable surveyors Edward Jarvis and Charles Hanington through the northern Canadian Rockies in the winter of 1874, all the gruesome details of which are duly enumerated in Sir Sandford Fleming's 1876 Report to the Canadian Parliament."[9]

All in all, the collection of references over a period of 140 years is very slim. None is long or detailed. As far as we are aware, this is the first book devoted entirely to the expedition and to Jarvis and Hanington.

APPENDIX E

Mal de raquette

Most Canadians, even most Canadian physicians, may not know this malady, partly because snowshoeing has, in many parts, been superseded as a means of backcountry winter travel, and partly because today's snowshoes are substantially lighter and more comfortable than those of yesteryear. In 1875, neither the cross-country ski, nor the snowmobile, nor the motor vehicle had been invented, and knowledge of the "snowshoe sickness" would have been widespread. It plagued both Hanington and Jarvis.

Hanington variously wrote in different letters to his brother Edward:[1]

> Jarvis had to follow behind slowly as he is suffering from *mal de raquette*. He doesn't say much about it but when he takes to the broken track with a white face and set lips you may guess he is in pain.

> I got very bad with *mal de raquette* yesterday and cannot recommend it as a travelling companion to any one who has to travel every day and all day.

> Down-hill travelling is worse for *mal de raquette* than up-hill, though I didn't think so when we were climbing.

Jarvis, too, made a number of comments:[2]

> *[M]al de raquette* was forgotten, (though it is generally a pretty attentive companion).

> [T]he unfailing attendants on hard travel in winter – snow-blindness and "*mal de raquette*."

> But on the river itself the depth did not exceed two feet or two feet and a half; into this, however, the snow-shoe would sink a good foot, and coming up with a small avalanche on the toe at each step, caused many blisters and occasional *mal de raquette.*

Mal de raquette may affect beginners within hours, but it can also affect the fit and the experienced in the form of an overuse injury on gruelling, long-distance trips. It can affect the foot, ankle, calf or inner thigh.

Descriptions of the affliction are vivid and scary. One website provides painful detail:

> The French Canadians had a great name for it, when you went lame while using snowshoes. Mal de raquette must be experienced to be fully appreciated. The main muscle on the inside of each leg at the groin becomes over taxed with the constant lifting of each foot and snowshoe out of the deep snow in order to take the next stride. The nose of the snowshoe must be lifted high enough to clear the snow and this results in an exaggerated lift of the knee. The weight of the snowshoe and the snow on it, seems to get heavier and heavier with each mile trod. Eventually a person reaches the stage where all the will power cannot lift the snowshoe up out of the hole one more time. If you are miles from home and in deep snow you are now trapped. You are an invalid in the wilderness. Sometimes a rest of a few hours can improve the situation and if you are not alone the other person will break trail.[3]

APPENDIX F

Geographic Names

The Jarvis and Hanington names are attached to a variety of features in the heart of the Rockies where they crossed the pass (see Chapter 6). Subsequent travellers, who admired their accomplishments, recommended these sites.

Jarvis and Hanington added only two names to the map, for the prominent pyramidal peaks at each end of the pass. The lower of the two, at the east end, they named Smoky Peak, as it marked their entry into the Smoky River watershed. It is now known as Mount St. Andrew's. At the west entrance is the stunning, Matterhorn-like peak that they named Mount Ida – without explanation. There is no mention in Jarvis's report, nor in his diary, of the origin of the name, just the prosaic entry, "To the most prominent of these points we gave the name of 'Mount Ida.'"[1] Even 50 years later, Hanington, in his reminiscences, made no mention of the naming.

This has provided a field day for speculation: Why "Mount Ida?" No one seems to know. A subsequent explorer, Frederick Vreeland, wrote in 1930: "E.W. Jarvis, an intrepid explorer for the proposed C.P.R... described...a very conspicuous peak eight miles northeast of Mount Sir Alexander, and named it, for reasons best known to himself, 'Mount Ida.'"[2]

Others simply repeated the fact that Jarvis had named Mount Ida, for example, Samuel Prescott Fay: "a man named Jarvis had crossed from the Fraser waters to the Porcupine through this valley up which we had come, and to the knife-like mountain he had given the name of 'Mount Ida.'"[3]

A possible clue may lie in Jarvis's predilection, in his report, for classical Latin phrases that have their origin in Virgil's *Aeneid*. A scholar versed in the classics might want to name a magnificent peak after a mountain in Greco-Roman antiquity. Mount Ida is the highest summit on the Greek island of Crete, and has an impressive pedigree. It was sacred to the Titan goddess Rhea, and no less than Zeus, head of all the Greek gods, was said to have been born within a cave on its slopes, *Idaion Andron*.[4]

Alice and Fred Dunn, in describing their 1954 first ascent of Mount Ida, had a different theory: "To the northeast of the ancient city of Troy, indeed flanking and protecting the city, is a mountain known as Ida. Perhaps the C.P.R. explorer, Jarvis, had this in mind when he named the last consequential peak of the north Canadian Rockies, and perhaps no."[5]

Olive Knox, who in writing historical fiction in *The Young Surveyor* had access to Jarvis's diary, developed the following fanciful version:

> Jarvis pointed at the most prominent peak, separated from the others by a long glacier, its face a transparent hue in the moonlight.
>
> "Mount Ida, I'll call that one," he said, and quoted Tennyson [from the first verse of *Oenone*]:
>
> Here lies a vale in Ida, lovelier
>
> Than all the valleys of Ionian hills.
>
> The swimming vapour slopes athwart the glen,
>
> Puts forth an arm, and creeps from pine to pine,
>
> And loiters, slowly drawn[6]

Like the naming of another prominent mountain, Mount Robson, the reason for naming Mount Ida will remain an enigma – Jarvis and Hanington's little joke on subsequent generations.

How, then, did the place names in the region of the pass come to be officially adopted? Some are more straightforward than others. R.W. Jones, who surveyed the area for the Grand Trunk Pacific Railway in 1906–1910, first used the term "Jarvis Pass" in 1906. The Geographic Board of Canada officially adopted it in 1917.[7]

Jarvis Lakes was likewise initiated by Jones, and adopted in 1917, confirmed in 1929 and again approved in 1965.

The situation becomes more complex with regard to Mount Jarvis and Mount Hanington. The concept of having the big mountains that guard the eastern end of Jarvis Pass named after Jarvis and Hanington had the obvious attractions of symmetry and relevance. Jarvis and Hanington had called the southern mountain "Smoky Peak." Once it became evident that the head of the Smoky River was further to the south, this name was clearly inappropriate. Jones named it Mount Jarvis, and this name was officially adopted in 1917.[8]

However, in 1923, the influential Boundary Commissioner A.O. Wheeler decided to call it Mount St. Andrew's. His reason: the pastor of St. Andrew's Church, Ottawa, was visiting the area at the time and had suggested to Wheeler that the feature be named after his church. Things unravelled further. In 1929, British Columbia surveyor A.J. Campbell, unaware that the name referred to a church, assumed that "Mount St. Andrew's" was a direct reference to the patron saint of Scotland, and used "Mount St. George" and "Mount St. Patrick" after the patron saints of England and Ireland, respectively. Mount St. David, referring to the patron saint of Wales, was designated in 1958.[9] The result was a profusion of names of important summits that had no significance for the area.

Samuel Prescott Fay maintained a lifelong interest in the area his party had explored in 1914, and had both an appreciation of history (as evidenced by his brief correspondence with Hanington, noted in the Introduction) and a sense of fairness.[10] To confuse matters further, he had called the peak Mount Koona during his trip. Once he realized that Jones had used the term Mount Jarvis, he readily accepted this. It was he, who in 1915, in correspondence with the Geographic Board of Canada, asked for the name to be changed from Smoky Peak to Mount Jarvis. Over 40 years later, in 1957, he re-entered into correspondence with the board, pointing out the injustices that had been perpetrated with the Mount St. Andrew's issue. It was noted that as early as 1932 there had been general agreement to rectify this unfortunate decision but that this had been overlooked by the surveyor who produced the map of the area, which therefore remained "Mount St. Andrew's." The board was perplexed, and the ultimate result was an unsatisfactory compromise. Mount St. Andrew's survived as the name of the high peak, and Mount Jarvis was applied to a much lower, subsidiary summit on its northern spur.

Fay had a way with words, and his contribution in 1957 remains resonant today:

> It seems a great pity to have so many of these early, very appropriate names changed to names that often have no special historical significance. The way it was prior to this last change, Mount Hannington was on the north side and Mount Jarvis on the south side of the exact height of land of Jarvis Pass; nothing could have been more appropriate in commemorating the very remarkable survey exploration trip in the dead of winter in 1874–75 undertaken by these two men.[11]

By contrast, the naming of Mount Hanington was uncontroversial – it was adopted in 1925 and has stood the test of time. And the current Mount Jarvis, while not lofty or magnificent, at least still lies just across Jarvis Pass from Mount Hanington. Fay's wish did not entirely fail to come true.

APPENDIX G

Geographical Features

The portion of the northern Rockies that Jarvis and Hanington and their party crossed is of singular beauty. There is a nice symmetry to the discoveries and descriptions they provided: a fine waterfall on either side of the mountains, a pass that slices through in west-east precision, flanked to the north and south by high peaks (two of which are officially named Mount Jarvis and Mount Hanington) and a great spire that is the northernmost 10,000-foot summit in the Rockies. It has been 140 years since Jarvis and Hanington provided their first descriptions, and it is therefore interesting to consider how these features have fared in the interim.

Jarvis and Hanington provided the first written descriptions of Herrick Falls, west of the mountains, and Kakwa Falls, east of the mountains. Both provided their party with dramatic moments.

Herrick Falls today can be visited either by jet boat or by means of a short hiking trail that leads off the Herrick Forest Service Road. It is a powerful place, with the roaring water plummeting down a near-vertical bedding plane and then hitting a headwall and being forced to make an abrupt 90-degree turn to the right. Regrettably, the condition of this road can never be assumed to be good.[1]

Kakwa Falls (see Figures 50 and 51), today in Alberta's Kakwa Wildland Park, is an adventure destination in itself, accessible via a long drive from Grande Prairie, the last arduous 20 kilometres of which involve creek crossings and passes, and require a 4WD truck, ATV or mountain bike. Kakwa Falls is sometimes touted as "Alberta's highest waterfall." When water levels are low, the river splits into two parts, both of which plunge uninterrupted into the great pool below. Perhaps the most interesting feature, and one not mentioned by Jarvis and Hanington (as they probably did not feel the need to venture back upstream to the base of the falls from the place where they careened into the canyon), is the large cave that exists behind the falls. Entering this great cavern and venturing into its dank, spray-filled recesses is a surreal experience, with the waterfall visible from behind.

Jarvis and Hanington were understandably preoccupied with survival. Their descriptions are succinct. The first, more detailed description was provided over 40 years later by a remarkable woman. Mary Jobe was one of the first female climbers in the Canadian Rockies. She would go on to great fame and decoration as a wildlife photographer and conservationist in central Africa. In the fall of 1917, she and her friend and guide Curly Phillips embarked on a trip to the headwaters of the Wapiti River. She described this expedition in the 1918 edition of the *Canadian Alpine Journal*, and included this evocative description of Kakwa (Porcupine) Falls:

> At the end of our fourteenth day of snowstorm, and eighteen days after leaving Mt Robson, we reached our final destination on the Wapiti, where in a spruce forest Phillips built a fine cabin for his supplies. There is no space here to tell of our return journey. I want merely to mention the splendid waterfall we explored on the Porcupine. Only a few white men have seen it, and so far as I know, no other white woman has been in that locality. The water falls in two great streams from a height of 225 feet into a gigantic dark green pool, below which rapids rush through a canyon six miles long. Behind the falls is a great ice cavern, into which one may walk dry-shod save for the blowing spray. With their fine background of forested and snow-capped mountains, and their adjacent banks, lined with hoodoos, the Falls of the Porcupine present a spectacle never to be forgotten. Several falls of lesser magnitude are in the canyon.[2]

Jobe might not have known that Jarvis and Hanington had been the first to write about Kakwa Falls, but she knew enough of their heroic expedition to include this passage in an article in the 1916 *Canadian Alpine Journal*:

> We took the trail north to Jarvis Pass, where as early as 1875 E. W. Jarvis, in a long surveying expedition, in which he travelled

> from Ft. George to Lake St. Anne, found this pass between the Fraser and the Porcupine...He travelled over 900 miles on snow shoes. The last 300 miles he carried a pack, making never more than ten miles a day. His locating this pass was a splendid achievement.[3]

Mount Ida was the most impressive phenomenon they encountered. In this region, only Mount Sir Alexander rises higher, but it is a massif; Mount Ida, by contrast, is pyramidal, tapering relentlessly to a pointed summit. In the early 20th century, it became clear that there were these two great mountains in the northern Rockies, the northernmost 10,000-foot peaks in the range.

Mount Sir Alexander (10,720 feet, or 3270 metres above sea level) was summited first, in 1929, probably attracting attention because of its greater bulk. Mount Ida (10,472 feet, or 3189 metres above sea level) was not summited until 1954. Alice and Fred Dunn reported on this achievement in the 1955 edition of the *Canadian Alpine Journal*.[4]

Other early explorers remarked on Mount Ida, Jarvis Pass and Jarvis Lakes. After Jarvis and Hanington, the next, chronologically, was R.W. Jones. His maps (including Jarvis Pass) have been preserved, but unfortunately his journals have not been traced. Then Samuel Prescott Fay penned these vivid lines in his 1914 journal:

> Now and then we would come to a little lake, some of them a mile or two long and of the most exquisite emerald green I ever saw. The contrast of the colour with the dark green of the spruces along the shores, in turn contrasted with the lighter green where the sunlight filtered through to the underbrush beneath – all these as a foreground to the massive peaks of sheer rock and ice beyond was a sight not easily forgotten.
>
> Soon we were above timberline and on a fine large open summit with several basins and one or two little lakes. As we turned around there lay a sight that not one of us will ever forget and which more than made up for the seven and a half hours on the trail. Across the deep valley we had come from rose a high pyramidal mountain, fully 11,000 feet, with nothing but sheer, unscalable cliffs on every side we could see...This is without doubt the most magnificent, the grandest view we shall ever have from a camp and it is something worth going thousands of miles to see. The mountain south of us

> is a splendid replica of the Matterhorn, and I think nearly as inspiring.[5]

Fay thought Mount Ida was unclimbable. He awoke in the morning of August 12, 1914, at what he called "Matterhorn Camp," then "before breakfast I rushed around taking pictures of the peak from several viewpoints and getting two panoramas of three photographs each, one of which was taken across a little lake nearby with all the mountains reflected in it."[6]

The result was arguably the most spectacular series of photographs ever taken of Mount Ida (see Figure 52). His group spent the day rambling and climbing (including what may have been Mount Hanington). They looked down from above on "our string of six lakes, the longest over two miles and of the most exquisite emerald green."[7]

In the 1920s, H.F. Lambart wrote:

> From the air this pass is a striking feature, flanked as it is with its glacier-clad Mt Ida, which, viewed from the pass itself, might be termed the "Matterhorn of the North." Jarvis Pass, its slopes clad with splendid timber, is distinctly marked by a series of five small lakes. Peacefully shimmering in the clear air, they looked like the links of a silver chain.[8]

A.O. Wheeler and Lambart co-authored these similar lines in the *Canadian Alpine Journal* of 1923: "Jarvis Pass seems to consist of a long valley, the floor of which is made up of a series of lakes, or one large, elongated lake, probably four to six miles long, cut into by rounded points projecting from either side. The valley seems well timbered with spruce, with an entire absence of any hardwoods."[9]

These comments were made based on views from an aeroplane. The next explorer to write of his travels was Prentiss Gray in 1928. Gray was an inveterate namer, and coined the place name "Barbara Lakes" after his daughter. This name didn't endure, unlike the creek that enters Jarvis Creek from Gray Pass to the north, which is still known as Barbara Creek. Gray waxed lyrical about the lakes and the peak that towered above them: "We had found no such beauty spot in all our travel: four lovely lakes, the forest pressing close to the water's edge, the placid surface reflecting the needle spire of Mt Ida with mirror-like precision."[10]

Division and impassioned debate about the various potential routes through the mountains and to the ocean persisted in the years after the Jarvis-Hanington trip. Jarvis and Hanington declared their route to be impractical for a railway. This conclusion did not deter subsequent explorers and surveyors from examining this area of the northern Rockies more closely. Beginning in the early 20th century, there was much exploration for passes suitable for railways. Initially, little was published, possibly due to fear of competition. The most thorough survey was by R.W. Jones, working for the Grand Trunk Pacific Railway. During 1905 and 1906, Jones and his crews explored most of the major potential routes between the Yellowhead and Pine passes. His 1906 maps, never published and only recently rediscovered, show all the routes they traversed from the Pine Pass in the north to Jarvis Pass in the south, and identify a preferred rail line via the Pine Pass.

Even after the Canadian National Railway was built through the Yellowhead Pass, there still remained the issue of a rail outlet for the farmers of the Peace region of northwestern Alberta and northeastern British Columbia. In 1927, Prentiss Gray entered the area north of Jarvis Pass. He was a successful New York banker who spent his summers in diverse global locations to pursue his passions for hunting and photography. He wrote detailed journals for his sisters, which were lovingly preserved by his son Sherman, and eventually published as a coffee table book: *From the Peace to the Fraser.* He claimed to have discovered what is still known as Gray Pass, one pass north of Jarvis Pass. In 1928, he guided a railway engineer (H.G. Dimsdale) through it, and proved the feasibility of the route for a railway. West of the divide, this route led down Barbara Creek (named after Gray's daughter) to Jarvis Creek, then via the McGregor and Torpy rivers to Bend on the Canadian National Railway line. (Jarvis and Hanington would have encountered the confluence of Barbara Creek with Jarvis Creek on their way up to Jarvis Lakes.) Had this become the rail outlet for the Peace Region, Mount Sir Alexander and Mount Ida might nowadays rival Jasper and Banff in the public perception of the Canadian Rockies.

Fortunately, for all who love their wilderness preserved, Jarvis Pass, Jarvis Lakes, Mount Ida, Mount Jarvis and Mount Hanington are all protected in their natural state as the explorers saw them, within British Columbia's Kakwa Provincial Park, proclaimed in 1999 and comprising over 170,000 hectares. To the east, over the interprovincial boundary, the park is contiguous with the 16,000-hectare Kakwa Wildland Park in Alberta, proclaimed in 1996. Hence, the route that Jarvis and Hanington pioneered through these mountains in 1875 is fully protected in a natural state.

NOTES

INTRODUCTION

1. Marcus Smith, "Special Report on Passes through the Cascade and Rocky Mountain Chains," in *Canadian Pacific Railway: Report of Progress on the Explorations and Surveys up to January 1874*, Sandford Fleming (Ottawa: MacLean, Roger & Co., 1874), 216–217.
2. Gerry Andrews, *Metis Outpost: memoirs of the first schoolmaster at the Metis settlement of Kelly Lake, BC, 1923–1925* (Victoria, BC, 1985).
3. Samuel Prescott Fay correspondence, Fred Brewster fonds, M53/V86, Whyte Museum of the Canadian Rockies, Banff, Alberta.
4. C.F. Hanington, "Journal of Mr. C.F. Hanington from Quesnelle through the Rocky Mountains, during the Winter of 1874–5," in Dominion of Canada, Parliamentary Sessional Papers, Vol. 5, Second Session of Sixth Parliament, Ottawa, 1888. Note C in Report on Canadian Archives 1887 by Douglas Brymner, Archivist.

CHAPTER 1
E.W. JARVIS'S REPORT AND NARRATIVE

1. *Report on Surveys and Preliminary Operations of the Canadian Pacific Railway up to January 1877*, by Sandford Fleming, Engineer in Chief. Ottawa, 1877. Appendix H. Reprinted with the permission of Library and Archives Canada. AMICUS 2645036/P.145–161.
2. These were the headwaters of a tributary of the Smoky River, the Kakwa River, therefore not what is now known as the Smoky River. The term "Smoky River Pass" was therefore inappropriate and did not endure.
3. This origin of the name, reported by the Cree, is provided also in *Place Names of Alberta*, vol. 4, *Northern Alberta* (Calgary: University of Calgary Press, 1996), 196.
4. Simon Fraser, in 1808, used the term "Quesnel's River" after one of the two clerks of the Northwest Company who had accompanied him on his historic journey to the mouth of the Fraser River. The town that developed at the junction of this Quesnel River with the Fraser River also came to be known as Quesnel. It has sometimes been misspelled as Quesnelle. See Helen B. Akrigg and G.P.V. Akrigg, *British Columbia Place Names* (Vancouver: University of British Columbia Press, 1997)

and BC Geographical Names, http://apps.gov.B.C..ca/pub/B.C.gnws/names/22304.html.

5 The Collins Overland Telegraph was an ambitious attempt from 1865 to 1867 to lay a telegraph line from San Francisco to Moscow, avoiding the need for long undersea cables across the Atlantic. Work on the section that Jarvis and Hanington used was done in 1866, with telegraph line installed over 600 kilometres. The project was abandoned in 1867 when undersea cable technology became viable. Hazelton, Telkwa and Telegraph Creek received their first European settlers as a result. Parts of the old telegraph wires may still be seen in places.

6 The Blackwater River is a major tributary of the Fraser River. It rises in the Ilgachuz Range, northwest of Quesnel, and flows eastwards, through a series of canyons, before joining the Fraser River.

7 H.P. Bell was a civil engineer also working on surveys for Canadian Pacific Railway (CPR). In 1874, two crews had surveyed the upper Fraser River between Tête Jaune Cache and Fort George. Jarvis's '*M*' crew had started at Tête Jaune Cache and worked west, while Bell's *N* crew worked east from Fort George. Jarvis (and Sandford Fleming) would therefore have been well acquainted with him. Alfred R.C. Selwyn, "Report on Exploration in British Columbia," in *Geological Survey of Canada: Report of Progress 1875–1876* (Montreal: Dawson Brothers, 1877), 28–86; The Early Years of the CPR in BC, http://canyon.alanmacek.com/index.php/Surveying#1874.

8 The Northwest Company fur trading post of Fort George was established by Simon Fraser in 1807 at the forks of the Fraser and Nechako rivers and named in honour of King George III. The name was changed in 1915 when the city of "Prince George" was incorporated.

9 The Giscome Portage led for 15 kilometres between the Fraser River and Summit Lake, which connected with the Peace River, thus crossing the Continental Divide and connecting the Pacific and Arctic watersheds. It leaves the Fraser River just under 60 kilometres upstream from Prince George. It is today the site of a heritage trail, used by hikers and cross-country skiers. The portage was long used by the Lheidli T'enneh First Nation as a trade route but is named after a Jamaican gold prospector to whom they showed it. See BC Parks' Giscome Portage Trail Protected Area, http://www.env.gov.bc.ca/bc.parks/explore/parkpgs/giscome/.

10 What is now known as the McGregor River was referred to by Jarvis as the North Fork of the Fraser. This river, in turn, forks into what are now known as Herrick Creek (known to Jarvis and Hanington as the North Branch of the North Fork) and the McGregor River (Jarvis and

Hanington's South Branch of the North Fork). The McGregor River has also sometimes been referred to as the Big Salmon River. See BC Parks' McGregor River, http://apps.gov.B.C..ca/pub/B.C.gnws/names/19543.html.

11 "Map of the Country Within the Rocky Mountain Zone," Sheet No. 8, in *Canadian Pacific Railway: Report of Progress on the Explorations and Surveys up to January 1874*, Sandford Fleming (Ottawa: MacLean, Roger & Co., 1874).

12 Herrick Creek.

13 McGregor River.

14 Jarvis Lakes.

15 The Athabasca River is a major tributary of the Mackenzie River. It arises in the Athabasca Glacier in Jasper National Park, and flows past Jasper. Its valley formed a natural destination for Jarvis, as from it he would have been able to head east along a route that Fleming had requested he explore.

16 Kakwa River.

17 The Fiddle River depot was built by a Mr. McCord, upon instructions from Walter Moberly (who, in turn, reported to Sandford Fleming), as the supply station at the eastern end of the Yellowhead Pass. McCord began building the depot on December 30, 1872. It would be used to provision the CPR trail and survey crews in the following years. Sandford Fleming, *Canadian Pacific Railway: Report of Progress on the Explorations and Surveys up to January 1874* (Ottawa: MacLean, Roger & Co., 1874), 170.

18 Walter Moberly was one of the larger-than-life characters who embellish the annals of the CPR surveys in the mountains. To quote Pierre Berton, he was "egotistical, impulsive, stubborn and independent of spirit. He could not work with anyone he disagreed with; and he disagreed with anyone who believed there was any other railway route to the Pacific than the one that had been developing in his mind for years." Pierre Berton, *The National Dream: The Great Railway, 1871–1881* (Toronto: McClelland and Stewart Ltd., 1970), 159. His route was Eagle Pass and Howse Pass. Fleming ordered him to abandon this in favour of the Yellowhead, at great inconvenience to his crews. The two men met there during Fleming's trip west, and Moberly received a dressing down. A few months later, reluctantly reconciled to the likelihood of the Yellowhead route, he ordered the building of the Fiddle River depot.

19 In 1813, the Northwest Company built Rocky Mountain House on the shores of Lac à Brûlé (close to where the Jarvis party struck the Athabasca valley); the name was later changed to Jasper's House after one of its factors, Jasper Hawes. In 1829, it was moved upstream to a site on Jasper Lake, 35 kilometres east of present-day Jasper. It served an important role in the fur trade. It was officially closed by the Hudson's Bay Company in 1884; by 1875, its importance had already dwindled. See Parks Canada, Jasper National Park, http://www.pc.gc.ca/eng/pn-np/ab/jasper/natcul/natcul11/d.aspx.

20 Beginning in 1795, there was a series of Northwest Company and Hudson's Bay Company posts in the approximate region of what is now the city of Edmonton. The site that Jarvis passed through was established by the Hudson's Bay Company in 1830, at the site of the present-day Alberta legislative buildings. Fort Edmonton was the western terminus of the Carlton Trail and thus the end of the main route west from the Red River Settlement and Fort Garry (now Winnipeg). Fort Edmonton Park has been created nearby as a tourist and historic attraction.

21 Lac Ste. Anne was settled by Métis in the 1830s. The Mission of Lake St. Anne was created in 1842. It grew and for a time enjoyed a greater population and commerce than Fort Edmonton (65 kilometres to the southeast). It would become a haven for disenchanted Métis moving west after the 1885 Northwest Rebellion. In time it became a famous Catholic mission, with a lake that reputedly had healing powers that to this day serves as a destination for pilgrimages and gatherings. The community of Lac Ste. Anne acquired a Hudson's Bay Company store in 1861, and later a North West Mounted Police barracks. It was the company store that was the attraction for Jarvis.

22 After completing the surveys through the Yellowhead Pass and surrounding Rockies, Walter Moberly had turned his attention to the most suitable route east from the Fiddle River depot.

23 The precise location of Root River is unclear. Fort White Earth was situated on the banks of the North Saskatchewan River over a hundred kilometres northeast of Edmonton. It operated from 1810 to 1813, and its palisade enclosed both Northwest Company and Hudson's Bay Company structures, for mutual protection. A line from Root River to Fort White Earth would presumably have been to the north of the route Jarvis chose to follow to Lake St. Anne and Edmonton.

24 Fort Victoria, also known simply as Victoria or Victoria Settlement, was established on the north bank of the North Saskatchewan River as a fur trading post of the Hudson's Bay Company in 1864. It played a role in

the settlement of eastern Alberta. The site is preserved as a museum and includes the oldest building in the province.

25 Fort Pitt was a Hudson's Bay Company trading post built above the north bank of the North Saskatchewan River in 1830. It is located just east of the current Alberta – Saskatchewan border. It was the major post between Fort Edmonton and Fort Carlton. It was the scene of the Battle of Fort Pitt during the Northwest Rebellion of 1885. It now forms Saskatchewan's Fort Pitt Provincial Park.

26 Fort Carlton, 65 kilometres north of present-day Saskatoon, was a Hudson's Bay Company fur trade post from 1810 until 1885. It was situated on the south bank of the North Saskatchewan River at the point where the Carlton Trail crossed it, and it became the central point on this thousand-mile trail from Fort Garry to Fort Edmonton. It was also the meeting point between this east-west route and a route north to Churchill River country. It is now a Saskatchewan provincial historic park. See Fort Carlton, *Wikipedia*, http://en.wikipedia.org/wiki/Fort_Carlton; Victor Carl Friesen, *Where the River Runs – Stories of the Saskatchewan and the People Drawn to Its Shores* (Calgary: Fifth House Press, 2001).

27 See Appendix E: *Mal de raquette* for details on this affliction.

28 The introductory paragraph by Sandford Fleming and the narrative were published in Fleming's *Report on Surveys and Preliminary Operations*.

29 "Mr. Bovil," according to Hanington, was a son of the chief justice of England, and was in charge of Fort George. Sir William Bovill was an English lawyer, Member of Parliament and judge. He served as Chief Justice of the Common Pleas between 1866 and his death in 1873. He had three sons; the most likely to have been in Fort George in 1874 was Eustace Bovill (1846–1912) who never married. See Halhed Geneaology & Family Trees, http://www.halhed.com/t4r/getperson.php?personID=I3294&tree=tree1.

30 Cottonwood Canyon is a canyon along the Fraser River northwest of Quesnel, downstream from its confluence with the east-flowing West Road River and upstream from its confluence with the northwest-flowing Cottonwood River. The first European explorer was Simon Fraser who ran the canyon's rapids in 1807. It was a recognized obstacle for steamboats operating on the Fraser River from Quesnel to Fort George. See Cottonwood Canyon (Fraser River), *Wikipedia*, http://en.wikipedia.org/wiki/Cottonwood_Canyon_(Fraser_River).

31 "Doubling up" refers to carrying the loads in two trips over each section of the trail because there is too much for one continuous trip.

32 Fort George Canyon is on the Fraser River south of Prince George. Up until 1914, this river provided the main access to northern British Columbia. Fort George Canyon, with its rapids, whirlpools and submerged rocks, was one of the obstacles that steamboats had to navigate. The area is now protected as a British Columbia Provincial Park.

33 Stuart Lake, referred to as "Stuart's Lake" by Jarvis, on the southeastern shores of which was situated Fort St. James, was often known in the 19th century as Stuart Lake Post. Fort St. James was founded by Simon Fraser in 1806. The fort is a National Historic Site of Canada (www.pc.gc.ca/eng/lhn-nhs/B.C./stjames/index.aspx).

34 A surveyor's chain is 66 feet (22 yards, or 100 links, or 4 rods), used as a measure of length in Britain for centuries. An engineer's chain, sometimes used in North America, is 100 feet. "Forty paces to the chain," when walking in snow on snowshoes, indicates that the surveyor's chain of 66 feet was used by Jarvis.

35 McGregor River.

36 What was formerly known as the Bear River is now the Bowron River, with its source in the Bowron Lakes. It was named after the gold commissioner in Barkerville, John Bowron.

37 This is the first written description of Herrick Falls.

38 Coign of vantage: a somewhat obsolete term that refers to "an advantageous position or stance for observation or action." *The Free Dictionary*, www.thefreedictionary.com/coign+of+vantage.

39 The term "Pullman car" refers to the railroad sleeping cars, renowned for their comfort and the views afforded from them, built by the Pullman Company.

40 "The Chief" was Jarvis's term of respect for Sandford Fleming.

41 The "south fork of the north branch of the Fraser River" is the McGregor River.

42 This is the Canyon of the McGregor, or Three Mile Canyon, so named because it lies three miles upstream from the junction of the McGregor River and Herrick Creek, although coincidentally Jarvis refers to a three-mile detour to get around it.

43 York Factory was a settlement and trading post at the mouth of the Hayes River on the southwestern shores of Hudson's Bay, and was a major post of the Hudson's Bay Company. It is a Canadian National Historic Site.

44 Jarvis refers to Mount Ida as one of the pyramidal peaks at the entrance so this would eliminate the western peaks, Dimsdale and Nechamus.

More likely the other is Mount Hanington seen end-on – it is a long ridge with numerous summits.

45 The glacier Jarvis refers to is most likely the Cheguin Icefall, on the south side of the pass, that dominates the local landscape.

46 Jarvis appears to be describing here the westernmost of the Jarvis Lakes.

47 The Jarvis Lakes.

48 This is the first written description of Jarvis Pass.

49 The blazing of the tree was done at the Continental Divide. The boundary with the Northwest Territory was a further six miles to the east, on the 120th meridian. Even in 1874–1875 the boundaries of British Columbia, established in 1866, were similar to today.

50 Smoky Peak likely refers to Mount St. Andrew's; Mount Hanington is the other contender. Neither Jarvis nor Hanington specified if the peak guarding the eastern end of the pass was on the north or south side of the pass.

51 Jarvis was quoting Virgil in *The Aeneid* (6.126). Literally translated as "the descent of Avernus is easy," *facilis descensus Averni* also has the connotation of "it is easy to slip into moral ruin," or "the road to evil is easy." Avernus was a deep lake, reputedly an entrance to the underworld.

52 It appears that they missed the creek from Kakwa Lake coming in from the south.

53 This is the first written description of Kakwa Falls.

54 The Northwest Company established Fort Dunvegan on the Peace River in 1805. In 1875, it was still the major outpost in the region.

55 The Hudson's Bay Company established Fort Assiniboine on the Athabasca River in the early 1820s. It became the western end of an 80-mile horse track to Fort Edmonton. It operated until the 1880s.

56 See Appendix A: The Smoky River Cache.

57 It is possible this was Prairie Creek.

58 Jarvis was under the impression they had descended the main branch of the Smoky River and that this was a southern branch or tributary.

59 These "high bluffs" are familiar to anyone who has driven Highway 40 between Grande Prairie and Grande Cache in Alberta. In the area where the Jarvis party crossed the Smoky River, the highway runs along the northwestern bank of the river for many kilometres, providing a view across the river of the high banks on the southeastern side, with few breaks. This creek up which they turned is most probably the Muskeg River or Wanyandie Creek.

60 The McLeod River (spelled "MacLeod" by Jarvis) is the large watercourse to the east of the Athabasca River that roughly parallels it from Hinton to Whitecourt, Alberta, where the two rivers join.

61 These names do not correlate with the current Baptiste River (north of Nordegg and Rocky Mountain House) or the Oldman River in southern Alberta.

62 Roche Miette has one of the most distinctive profiles of a Canadian Rockies peak, its thousand-metre vertical north face dominating approaches to the Yellowhead Pass from the east via the Athabasca River. It was named either after a colourful (maybe mythological) local named Miette, or else a derivation of the Cree name for the bighorn sheep that are found on its lower slopes.

63 The summit of Roche Miette is just five kilometres from where Jasper House was situated at the north end of Jasper Lake.

64 The southernmost of this series of lakes is Jarvis Lake. The Visitor Centre at William A. Switzer Provincial Park nestles on its shores at Kelly's Bathtub. Nearby is the large Jarvis Lake campground. The lake feeds Jarvis Creek. Little could Jarvis have imagined that his name would thus be remembered.

65 Lac à Brûlé is, as Jarvis suggests, a place of mighty winds funnelling down the Athabasca River. Like Jasper Lake, it is a shallow widening of the Athabasca River. The original Jasper House was built on its shores before being moved upstream to Jasper Lake.

66 See note 17 in this chapter.

67 Once again, Jarvis was quoting Virgil's *Aeneid* (3.57): *auri sacra fames* translates to the accursed greed for gold, or the holy lust for gold, implying that money is the root of all evil.

68 Walter Moberly had supervised some surveying east of the Rockies from the Yellowhead Pass in 1873, before leaving the Canadian Pacific Survey for good.

69 "Brûlés," as Jarvis uses the term, are areas of burned woodland.

70 This is the trail between the Athabasca River and the McLeod River.

71 Hanington noted that Dirt Lake was also known as Chip Lake, by which name it is still known. It is a large lake that drains eventually into the Pembina River. The Yellowhead Highway runs along its southern shore.

72 Despite "Mr. McGillivray" being the person whom Jarvis and Hanington were probably the happiest to see during their entire trip, exactly who he was is not clear, although McGillivray is a well-known name in the

annals of the fur trade. Canada's Historic Places, Alberta Register of Historic Places, River Lot 3, Victoria Settlement, https://hermis.alberta.ca/ARHP/Details.aspx?DeptID=1&ObjectID=4665-0971.

73 Richard Charles Hardisty (1831–1889) was the chief factor at Fort Edmonton from 1862 to 1864 and 1872–1883. He was appointed to the Canadian Senate on the advice of John A. Macdonald in 1888. The village of Hardisty, Alberta, is named after him. Richard Hardisty, *Wikipedia*, http://en.wikipedia.org/wiki/Richard_Hardisty.

74 Stuart Lake, Fort St. James.

75 In 1875, Tête Jaune Cache was the community at the western end of the Yellowhead Pass. (*Tête jaune* is French for "yellow head.") It was named after Pierre Bostonais, a blonde-haired Métis fur trader and trapper.

76 Fort Garry was the Hudson's Bay Company trading post at what is now downtown Winnipeg. It was established in 1822, and was the hub of the company's administration in the Red River Colony. The city of Winnipeg was established in 1873, and the new term was increasingly used at the expense of Fort Garry. The site is a National Historic Site of Canada.

77 In Jack (John) Norris they had an expert guide. Born in Scotland, John Norris (1826–1916) joined the Hudson's Bay Company and worked as a labourer and boatman. He helped lead the first brigade of Red River ox carts from Winnipeg to Edmonton in the early 1860s, a journey that took three and a half months, thus pioneering the route for others to follow. The Edmonton Historical Board, "Honouring Heritage Achievement," *Building Heritage* 2, no. 2 (December 1998): 2, http://www.edmonton.ca/city_government/documents/buildingheritagenews_vol_2_no_2_dec_1998.pdf.

78 Charles Adams was the clerk at the Hudson's Bay Company post at Fort Victoria in 1875. His descendants still live in the region. The house in which he stayed, the clerk's house, is the oldest building in Alberta that is still in its original location. It forms part of the Victoria Settlement National Historic Site of Canada (also a provincial historic site).

79 Dog Rump Creek enters the North Saskatchewan River from the north near the community of Elk Point, about 60 kilometres west of the Alberta – Saskatchewan boundary. This is not to be confused with the original name for Stony Plain west of Edmonton.

80 Having previously served at Fort Ellice, William McKay was the factor at Fort Pitt from 1873 until his death in 1882. He and his wife were known for their constructive relations with regional First Nations and Métis. Glenbow Museum, http://ww2.glenbow.org/search/

archivesPhotosResults.aspx?XC=/search/archivesPhotosResults.
aspx&TN=IMAGEBAN&AC=QBE_QUERY&RF=WebResults&DL=0&RL=0&
NP=255&MF=WPEngMsg.ini&MR=10&QB0=AND&QF0=File+number&Q
I0=NA-1193-5&DF=WebResultsDetails; Duck Lake Regional Interpretive Centre, Duck Lake, Saskatchewan, Chief Factors of the Hudson's Bay Company, http://www.museevirtuel-virtualmuseum.ca/sgc-cms/histoires_de_chez_nous-community_memories/pm_v2.php?id=hotspot_record_detail&fl=0&lg=English&cx=00000459&rd=115953&hs=1&alt=Factor+William+McKay+from+Fort+Pitt%2C+Saskatchewan; Gabriel Dumont Institute of Native Studies and Applied Research, http://www.metismuseum.ca/media/document.php/11394.William%20McKay%20II.pdf.

81 Lawrence Clarke (1832–1890) was the chief factor for the Hudson's Bay Company at Fort Carlton. He later became a magistrate, and went on to become the first elected member of the Legislative Assembly in the Northwest. Lawrence Clarke (politician), *Wikipedia*, http://en.wikipedia.org/wiki/Lawrence_Clarke_(politician); Stanley Gordon, "CLARKE, LAWRENCE," in *Dictionary of Canadian Biography*, vol. 11, University of Toronto/Université Laval, 2003–, http://www.biographi.ca/en/bio/clarke_lawrence_11E.html; Friesen, *Where the River Runs.*

82 Fort de la Corne was one of the two French forts established on the Saskatchewan River in the mid-18th century. Fort de la Corne, *Wikipedia*, http://en.wikipedia.org/wiki/Fort_de_la_Corne.

83 Cumberland House on the Saskatchewan River is the oldest permanent settlement in Saskatchewan and Western Canada. It was established by Samuel Hearne in 1774 as the first inland trading post of the Hudson's Bay Company. It was a vital supply depot for the fur trade era. The site is now part of Cumberland House Provincial Park.

84 The city of Prince Albert began as a mission post in 1866.

85 Fort Ellice.

86 South Saskatchewan River.

87 Mount Carmel lies six kilometres north of the community of Carmel, and west of Humboldt. By the time Jarvis passed this landmark, it had already been identified as "Mt Carmel" in a Geological Survey of Canada report of 1873–1874. It was known as *spathanaw watchi* in Cree, meaning "the mount of the far view," alternatively Keespitanow Hill, and was also known as Round Hill or the Big Butte or the Hill of the Cross. It is now the site of a Catholic shrine, built in 1937, dedicated to the Virgin Mary in her role as Our Lady of Mount Carmel (the original Mount Carmel is in Palestine). Its intriguing history is described in W.P. Telfer's *Humboldt*

on the Carlton Trail (Saskatoon: Modern Press, 1975), 35-37. Three years after Jarvis passed this way, it was the site of the suicide of a 19-year-old girl. It was subsequently alleged that the cross was erected as a memorial to this tragic event. Jarvis's comments that there was already a cross on the summit in 1875 were used to refute this claim. Mount Carmel is one of the most prominent and highest points in Saskatchewan. Perhaps the most evocative description ever written is by Sir William Butler, in his 1873 book *The Wild North Land*: "The hill of the Wolverine and the lonely Spathanaw Watchi have witnessed many a deed of Indian daring and Indian perfidy...Alone in a vast waste, the Spathanaw Watchi lifts his head; thickets and lakes are at its base, a lonely grave at its top, around four hundred miles of horizon; a view so vast that endless space seems for once to find embodiment and at a single glance the eye is satisfied with immensity." Rootsweb, Metisgen L Archives, http://archiver.rootsweb.ancestry.com/th/read/MÉTIS GEN/2002-02/1013057139.

88 South Saskatchewan River.

89 Gabriel Dumont's sign carried inscriptions in English, French and Cree. Dumont ran a small store, billiards hall and ferry service over the South Saskatchewan River in the late 1870s and early 1880s at what came to be called Gabriel's Crossing. The river is now crossed by the Dumont Bridge, east of Rosthern. It was built close to the original crossing site. Gabriel Dumont (1837–1906) was a renowned leader of the Métis people. In 1873, he was elected president of the St. Laurent republic. He invited Louis Riel back to Canada to help focus attention on the plight of the Métis people. He was adjutant general in the provisional Métis government declared in the District of Saskatchewan in 1885, and commanded the Métis forces in the Northwest Rebellion of 1885. He proved himself an astute military commander in these hostilities. Following defeat at Batoche, Dumont made for Montana and became a political refugee. He was a star attraction for a while in the Wild West Show of Buffalo Bill (William Frederick Cody) before being allowed to return to Canada, where he continued to serve as an advocate for his people. Gabriel Dumont (Métis leader), *Wikipedia*, http://en.wikipedia.org/wiki/Gabriel_Dumont_(M%C3%A9tis_leader).

90 Touchwood Hills Post was a Hudson's Bay Company trading post in Saskatchewan from 1852 to 1909. Unusually for such a post, it was not situated on a waterway. It formed a resupply point and convenient stopping place on the Carlton Trail. Touchwood Hills Post was designated a provincial park in 1986. Here, 16 kilometres east of Punnichy, a cellar depression and a segment of the Carlton Trail are preserved and erstwhile buildings are outlined. This is not the site

Jarvis passed, though, as the post was relocated by a few kilometres in 1879.

91 The Qu'Appelle River flows east for over 400 kilometres in southern Saskatchewan. Its waters are linked with those of the South Saskatchewan River through Lake Diefenbaker. Its valley dwarfs the river – it was once the route followed by the South Saskatchewan, draining the immense melting ice sheets. It joins the Assiniboine River just inside Manitoba near the village of St. Lazare and close to the historic Fort Ellice.

92 Fort Ellice was a Hudson's Bay Company trading post built in 1831 near the junction of the Assiniboine and Qu'Appelle rivers. It formed a major stopping point on the Carlton Trail. It was a staging point for part of the North West Mounted Police force that headed west in 1874 to establish law and order. It was named after Edward Ellice, a British merchant and an investor in the Hudson's Bay Company. As early as 1931 it was noted that just about nothing remained of this once bustling hub of activity. In contrast to almost all the other posts and forts that provided relief to Jarvis and Hanington, which have become provincial or national historic sites or parks, history has not been as kind to Fort Ellice. It was sold into private hands in 1925, and is currently a cow pasture. Recent attempts to repurchase it and develop it as a historic site have failed. See Bill Redekop, "The Battle of Fort Ellice," *Winnipeg Free Press*, January 30, 2010, http://www.winnipegfreepress.com/opinion/fyi/the-battle-of-fort-ellice-83137277.html.

93 There are numerous words of praise to be found for Archibald McDonald, a Scot born in 1836 who sailed to Canada on the Hudson's Bay Company vessel in 1854. He served the company at a variety of western posts and forts before becoming district factor at Fort Ellice from 1873 to 1879. In 1890, he suffered a fall off a democrat wagon along with fellow factor Richard Hardisty. Hardisty died from his injuries and McDonald suffered spinal injuries, the effect of which afflicted him for the rest of his life. He died in Fort Qu'Appelle in 1915.

One of the moving tributes was penned by Reverend A.B. Baird in a presentation to the Manitoba Historical Society in 1931. Baird had paused at Fort Ellice on an adventurous journey west in 1881.

> Archibald McDonald was a typical officer of the Hudson's Bay Company. He was naturally a high spirited and courageous man, and his fearlessness made him friends with the brave Crees of the plains with whom he maintained a friendship, and established an influence which prevailed through thick and thin through two rebellions, till his death. He was the last survivor in the active

service among the old chief factors of the Hudson's Bay Company. A picturesque frontiersman and pioneer, proud of the traditions of the Company and himself, doing much to advance its interest and its reputation.

Hudson's Bay Company Archives, HBCA 1987/363-E-700-MC/46, N15350, http://www.gov.mb.ca/chc/archives/hbca/; The Manitoba Historical Society, Memories of Fort Ellice, http://www.mhs.mb.ca/docs/mb_history/54/fortellice.shtml; Manitoba Historical Society, Archibald McDonald (1836–1915), http://www.mhs.mb.ca/docs/people/mcdonald_a3.shtml.

94 The Assiniboine River is the major river of southwestern Manitoba, over 1000 kilometres in length. It rises in Saskatchewan and flows south, often close to the Saskatchewan – Manitoba border before turning east and eventually flowing into the Red River in Winnipeg. It is joined by the Qu'Appelle River close to where Fort Ellice once stood, near the village of St. Lazare. Little could Jarvis know as he crossed it that one day he would supervise the construction of a bridge over it in Winnipeg.

95 In 1875, the North West Mounted Police sent a detachment to monitor traffic along the Carlton Trail. It established a barracks, stable and guardhouse beside a shallow lake known as Shoal Lake. Settlement followed and the community of Shoal Lake developed. See Shoal Lake History, http://shoallakehistory.wordpress.com/snapshots/.

96 The Little Saskatchewan is a sizeable river, originating in Riding Mountain National Park at Lake Audy and flowing through southwestern Manitoba for over 100 kilometres until it joins the Assiniboine River west of Brandon.

97 The Whitemud River is a meandering creek that begins near Neepawa, Manitoba, and ends in Lake Manitoba at Lynch's Point.

98 Portage la Prairie is now a small city located 75 kilometres west of Winnipeg, on the banks of the Assiniboine River. The first settlers came in the 1850s, attracted by the fertile plains. By the 1860s, there were 60 homes, and rapid settlement occurred in the 1870s. By the 1880s, the population reached 3,000; it was therefore the first sizeable community that Jarvis and Hanington reached since leaving Quesnel.

99 The flat area west of Winnipeg was known as the White Horse Plain, now occupied by the village of St. Francis Xavier. The name, and the legend of the white horse that engendered it, are ancient, dating back to the late 17th century.

100 H.A.F. MacLeod was, like Jarvis, one of the trusted surveyors that Sandford Fleming sent out west to investigate the lay of the land and its

suitability for a railway. MacLeod's summary was published as Appendix M in Sandford Fleming's *Report on Surveys and Preliminary Operations.* It is entitled "Progress report on the surveys made in the north-west territories during the year 1875, by H.A.F. MacLeod." The following year's report by Macleod appears as Appendix Y. Jarvis's report, in this same volume, is Appendix H. MacLeod's task was to survey the area from Fort Pelly (another important Hudson's Bay Company post in northeastern Saskatchewan, between present-day Kamsack and Pelly and now a National Historic Site) to the Yellowhead Pass. At the beginning of his report, MacLeod describes his trip west from Winnipeg but does not mention the chance encounter with Jarvis.

101 Sandford Fleming.

CHAPTER 2
C.F. HANINGTON'S JOURNAL

1 Dominion of Canada, Parliamentary Sessional Papers, Vol. 5, Second Session of Sixth Parliament, Ottawa, 1888. Note C in Report on Canadian Archives 1887 by Douglas Brymner, Archivist. Reprinted with the permission of Library and Archives Canada. AMICUS 1483203/P. cx-cxxxii.

2 The current spelling of the city name is "Quesnel." Using Hanington's letters, Leslie Middleton, of the Quesnel and District Museum and Archives, graciously researched and provided the information on Quesnel and its residents that appears in the notes that follow.

3 Subsequently referred to in this journal as "Hanington's cache."

4 The "North Fork of the Fraser" is now known as the McGregor River.

5 As becomes apparent, this is an incorrect term.

6 Prince George.

7 The two canyons on the Fraser River were Cottonwood Canyon and Fort George Canyon. Cottonwood Canyon is northwest of Quesnel; Fort George Canyon is south of Prince George.

8 Division M. was the survey party under the leadership of Jarvis.

9 Thomas Brown and Ben Gillis established the Occidental Hotel in 1865 to cater to the needs of miners on the Cariboo Waggon Road, and it was widely reputed to be the best stop on this trail. The Quesnel Hotel stands on the site of the old Occidental Hotel.

10 A Blue Noser is an inhabitant of the Maritimes, in particular Nova Scotia.

11 A Haligonian is someone residing in, or born in, Halifax.

12 There is a record of a Michael Haggarty (intestate) in the wills probated and letters of administration section, year 1897, of the Cariboo government agency records 1860–1938.

13 James Reid was a successful miner in the 1860s who then became an entrepreneur and invested in Quesnel. Over time he not only operated a large, well-stocked general store but also had interests in a sawmill, a flour mill, paddlewheelers and owned his own riverboat. He was elected the federal member for the Cariboo in 1881 and became a senator in 1888.

14 The *Cariboo Observer* of June 13, 1968, listed the 33 names on the Quesnel Voter's List of September 1, 1879, which included Girod, John, trader.

15 "By 1865 Kwong Lee and Company, formerly in mining with 'Billy' Ballou on Ferguson's Bar, but with mercantile interests in Victoria and Barkerville, had acquired their store on Barlow Avenue." Gordon R. Elliott, *Quesnel, Commercial Centre of the Cariboo Gold Rush* (Quesnel, BC: Cariboo Historical Society, 1958), 60.

16 "Cariboo" was a regional name first applied to the goldfield area around Quesnel and Barkerville but was extended to encompass the area between Cache Creek and Prince George. It was named for the animal, but it is not known for sure why it is spelled with a double "o" rather than "ou." Many believe it was because early inhabitants spelled phonetically.

17 The stagecoach was operated by the British Columbia Express Company, and the Cariboo Waggon Road was constructed in 1865 by the Royal Engineers to provide access to the goldfields.

18 James Pollock was a liveryman who pre-empted 160 acres (Lot 84) on the west bank of the Fraser River, four miles north of Quesnelle Mouth, in 1863. It appears that he was not running a stopping house but was simply good enough to give the travellers shelter for the night, as was the custom.

19 See note 5 in Chapter 1 on the Collins Overland Telegraph.

20 Jarvis noted "forty paces to the chain." As explained in note 34 in Chapter 1, this referred to the surveyor's chain, which was 66 feet long. This allowed for relatively accurate calculation of distance covered.

21 The Blackwater River is a major tributary of the Fraser River. It rises in the Ilgachuz Range, northwest of Quesnel, and flows eastwards, through a series of canyons, before joining the Fraser River.

22 Canadian Pacific Railway.

23 See note 29 on Mr. Bovil in Chapter 1.

24 Charles Ogden (1819–1890) joined the Hudson's Bay Company in 1853.

The next year, when at Fort Boise, he was accused of selling ammunition to local Indians. The result was a media call for the company to be evicted and its establishment burned. He was later transferred to British Columbia/New Caledonia and worked as a clerk in various posts. See BC Métis Mapping Research Project, http://document.bcmetiscitizen.ca.

25 Fort St. James, on the southeastern shores of Stuart Lake, was often known in the 19th century as Stuart Lake Post. Fort St. James was founded by Simon Fraser in 1806. The fort is a National Historic Site of Canada. Parks Canada, Fort St. James National Historic Site, www.pc.gc.ca/eng/lhn-nhs/B.C./stjames/index.aspx.

26 Hudson's Bay Company.

27 Hanington's cache was situated midway between where Herrick Creek joins the McGregor River, and the McGregor River joins the Fraser River.

28 Bowron River.

29 The Giscome Rapids form an area of turbulence on the Fraser River close to Giscome Portage.

30 McGregor River.

31 The Forks: the confluence of McGregor River and Herrick Creek.

32 Herrick Creek.

33 McGregor River.

34 Herrick Falls.

35 This long rapid is located at 54° 13' 01" north; 121° 06' 13" west.

36 This cascade and rock garden are situated at 54° 09' 01" north; 120° 49' 14" west.

37 On Hanington's 1888 map, No. 3 appears to be the upper Herrick flowing south from Herrick Pass.

38 The south branch appears to be Ovington Creek.

39 The Yellowhead Pass had been known to Europeans since 1820 and was identified by Sandford Fleming in 1871 as the preferred rail route through the Rockies.

40 McGregor River.

41 This is a puzzle – Three Mile Canyon is situated here on the McGregor River, but there are no substantial falls. There is talk of a massive logjam building up here and giving the impression of a drop that may have been more evident in winter and at low water. Wayne Giles, email correspondence.

42 This ghastly experience all relates to the portage around Three Mile Canyon on the McGregor River.

43 This could possibly be the forks of Jarvis Creek and McGregor River.

44 Jarvis.

45 This was a fine estimate. Herrick River and Jarvis Creek are just 17 kilometres apart at this point. Separating them is a broad, shallow, timbered valley, with a low pass that is just 140 metres (460 feet) above the level of Herrick Creek. Instead, to reach the point on Jarvis Creek from which Hanington made his comments, the party retreated and covered over 120 kilometres as the crow flies, and over 140 miles (225 kilometres) by their pacing measurements.

46 Jarvis recorded a 50-foot waterfall at this point.

47 Both Jarvis Creek and Edgegrain Creek (the southwest tributary mentioned in Jarvis's diary) drop precipitously through canyons near their confluence.

48 The pass is just below 5,000 feet and the peaks on each side are approximately 7,500 feet high, close to Jarvis's estimate; they evidently climbed just above treeline to obtain a clear view.

49 Jarvis Lakes.

50 The height above sea level of the highest of the Jarvis Lakes is 1472 metres (4,830 feet).

51 Hanington's "Smoky Peak" is probably Mount St. Andrew's.

52 This is the only mention of Mount Ida by Hanington.

53 They were on the Continental Divide. In this area, the boundary between British Columbia and the Northwest Territory (now Alberta) had been established along the 120th Meridian in 1866; it lies 12 km (7.5 miles) east of this point.

54 At this elevation in winter these birds were most likely ptarmigan.

55 There is no mention of the creek coming in from the south (from Kakwa Lake).

56 They were on what is now the Kakwa River, a tributary of the Smoky River.

57 Along with that of Jarvis, this is the first description of Kakwa Falls.

58 Lower Kakwa Falls.

59 The South Kakwa River.

60 See Appendix E for details on the affliction of snowshoe illness.

61 Possible candidates for this creek and the next creek mentioned (entering the Kakwa River from the south) include Daniel's Creek, Copton Creek or a series of unnamed creeks.

62 This creek is probably unnamed.

63 See Appendix A: The Smoky River Cache.

64 This was the valley of the Smoky River.

65 The Athabasca River is a major tributary of the Mackenzie River. It arises in the Athabasca Glacier in Jasper National Park and flows past Jasper. Its valley formed a natural destination for Jarvis, as from it he would have been able to head east along a route that Fleming had requested he explore. However, they were now on the Smoky River, not the Athabasca. Hanington had recently written that he was certain the river they had descended was the Smoky, whereas it was the Kakwa. Allowing for this, their wishful thinking that they had reached the Athabasca was understandable. If it had been the Athabasca, their troubles would have been minimal compared with what still lay ahead.

66 This tributary is most probably the Muskeg River. Wanyandie Creek is less likely.

67 See note 60 on the McLeod River in Chapter 1.

68 See note 21 on Lac Ste. Anne in Chapter 1.

69 See note 62 on Roche Miette in Chapter 1.

70 Lac à Brûlé and Jasper Lake are shallow widenings of the Athabasca River. The original Jasper House was built on the shores of Lac à Brûlé before being moved upstream to Jasper Lake.

71 See note 17 on the Fiddle River depot in Chapter 1.

72 See note 19 on Jasper House in Chapter 1.

73 The river they descended from the pass is now known as the Kakwa River; the "2nd one" is now named the Smoky River.

74 Hanington's account relates one version of the origin of Roche Miette. Another version is found in note 62 in Chapter 1.

75 See note 71 on Chip Lake in Chapter 1.

76 The Pembina River has its origin in the Rockies foothills south of Cadomin, Alberta. It flows for over 500 kilometres before joining the Athabasca River west of the city of Athabasca.

77 This would have been in the region of Lake Isle.

78 See note 72 on Mr. McGillivray in Chapter 1.

79 See note 76 on Fort Garry in Chapter 1.

80 See note 20 on the city of Edmonton in Chapter 1.

81 See note 25 on Fort Pitt in Chapter 1.

82 See note 24 on Victoria in Chapter 1.

83 See note 26 on Fort Carlton in Chapter 1.

84 See note 92 on Fort Ellice in Chapter 1.

85 See note 98 on Portage la Prairie in Chapter 1.

86 See note 77 on Jack Norris in Chapter 1.

87 Henry House was a depot built close to the site of present-day Jasper.

CHAPTER 3
EXCERPTS FROM E.W. JARVIS'S DIARIES, 1875

1 Jarvis's diaries are held at the Archives of Manitoba in Winnipeg. This material is from files MG6 – A2/5 and 6. The excerpts and sketches are published with the permission of the Archives of Manitoba.

2 According to Sam Steele, *Forty Years in Canada: Reminiscences of the Great Northwest, with Some Account of His Service in South Africa* (Toronto: McClelland, Goodchild, & Stewart, 1915), 88, Colonel W.D. Jarvis was a cousin of E.W. Jarvis.

CHAPTER 4
EXCERPTS FROM C.F. HANINGTON'S REMINISCENCES, 1928–1929, DESCRIBING THE 1875 EXPEDITION

1 Hanington's Reminiscences are held at the Royal BC Museum, BC Archives in Victoria, file E/E/H19. This excerpt is published with the permission of BC Archives.

CHAPTER 5
THE EXPEDITION IN PERSPECTIVE

1 E.E. Rich, ed., *Colin Robertson's correspondence Book, Sept. 1817 to Sept. 1822* (Toronto: The Champlain Society, 1939).

2 Smith, "Special Report on Passes"; A.C. Anderson, *Upper Fraser River: maps and notes submitted to Marcus Smith in support of a route for the Canadian Pacific Railway*, 1874, BC Archives CM/13699A. Smith's report is dated June 2, 1873, but the Anderson map at the BC Archives is dated January 13, 1874.

3 Henry John Moberly, *When Fur Was King* (London and Toronto: J.M. Dent, 1929).

4 Walter Moberly, *The Rocks and Rivers of British Columbia* (London: H. Blacklock and Co., 1885).

5 Selwyn, "Report on Exploration in British Columbia."

6 See the Historical Atlas of Canada Online Learning Project, http://www.historicalatlas.ca/website/hacolp/national_perspectives/exploration/UNIT_08_5/index.htm; Derek Hayes, *British Columbia: A New Historical Atlas* (Vancouver: Douglas & McIntyre, 2012).

7 See the map on page 129.

8 Smith, "Special Report on Passes"; Dominion of Canada, *Report of the Canadian Pacific Railway Royal Commission*, Volume 1 Evidence (Ottawa, 1882), 289.

9 Dominion of Canada, *Report of the Canadian Pacific Railway Royal Commission*, 289–290.

10 See Chapter 4.

11 Describing Smoky River, the website Spiral Road states, "This local name for the Morkill River was in use before the surveyor Dalby Morkill visited the area in 1913. Stanley Washburn camped on the 'Big Smoky' in 1909. It appears on the 1915 Provincial Pre-Emptors map as 'Morkill (Little Smoky).'" See http://www.spiralroad.com/smoky-river/.

12 See Chapter 4 of this book.

13 Dominion of Canada, *Report of the Canadian Pacific Railway Royal Commission*, 289.

14 For example, see E.W. Jarvis's diary entry for January 22, 1873 (Archives of Manitoba MG6 – A2/4) and for June 6, 1875 (Archives of Manitoba MG6 – A2/6).

15 Sandford Fleming's *Report on Surveys and Preliminary Operations*, Appendix F, written by Marcus Smith, mentions that in the summer of 1874: "Mr. Jarvis went up the Clearwater to the lakes, thence north-eastward across the divide to the Cariboo Fork of the North Thompson. The summit of the divide was fully 7,000 feet above sea level, at the lowest place he could find, which was over an immense glacier."

16 Excerpted from page 155 of *The National Dream: The Great Railway, 1871–1881* by Pierre Berton. Copyright © 1970 Pierre Berton Enterprises Ltd. Reprinted by permission of Anchor Canada/Doubleday Canada, a division of Random House of Canada Limited, a Penguin Random House Company.

17 "The Death of Major Jarvis," *Manitoba Free Press*, November 27, 1894, p. 4.

18 "Last Surveyor of C.P.R. Line Dies Aged 82 [obituary of C.F. Hanington],"

Winnipeg Tribune, December 22, 1930, p. 5, http://www.newspapers.com/newspage/39354205/.

19 G.M. Dawson, "Report on an exploration from Port Simpson on the Pacific coast to Edmonton on the Saskatchewan, embracing a portion of the northern part of British Columbia and the Peace River country," in *Geological and Natural History Survey of Canada: Report of Progress for 1879–80* (Montreal: Dawson Bros., 1881), 1–95.

20 Andrews, *Metis Outpost.*

21 Fleming, *Report on Surveys and Preliminary Operations*, 25.

22 Sandford Fleming, "Expeditions to the Pacific," in *Proceedings and Transactions of the Royal Society of Canada for the Year 1889*, Vol. 7, Section 2, Part 4 (Montreal: Dawson Brothers, 1890), 130.

23 See Chapter 1 of this book.

CHAPTER 6
BIOGRAPHIES

1 J.M. Bumsted, "The Household and Family of Edward Jarvis, 1828–1852," *Island Magazine*, no. 14 (Fall–Winter 1983), 22–28.

2 Archives of Manitoba, MG6 A2/1 – A2/9.

3 Olive Knox, *The Young Surveyor* (Toronto: Ryerson Press, 1956).

4 Baptismal record, Provincial Archives of PEI, http://www.gov.pe.ca/archives/parosearch/vital/individual-vital information/recordId/152452/eventType/1.

5 Bumsted, "The Household and Family of Edward Jarvis, 1828–1852," 22–28; https://thebravestcanadian.wordpress.com/2014/12/12/mystery-jarvis-gray-cousin-had-a-dramatic-life-of-accomplishments-in-the-canadian-frontier/.

6 Bumsted, "The Household and Family of Edward Jarvis, 1828–1852," 27, 28; J.M. Bumsted and H.T. Holman, "JARVIS, EDWARD JAMES," in *Dictionary of Canadian Biography*, vol. 8, University of Toronto/Université Laval, 2003 –, accessed May 8, 2015, http://www.biographi.ca/en/bio/jarvis_edward_james_8E.html.

7 Institution of Civil Engineers (ICE), Obituary for E.W. Jarvis: Minutes of the Proceedings, Vol. 124, Issue 1896, January 1896, pp. 239–240; Jarvis, diary entry for September 12, 1863, Archives of Manitoba.

8 Jarvis probate record, Archives of Manitoba, File 1685, Book H, Folio 41, http://www.gov.mb.ca/chc/archives/probate/wpg_estate.html#wpg1870;

Jarvis, diary entries for December 28, 1863, and January 13, 1864, Archives of Manitoba.

9 ICE, Obituary for E.W. Jarvis.

10 ICE, Election as associate member: Minutes of Proceedings, Vol. XXXVIII, Part II, Session 1873–1874.

11 ICE, Obituary for E.W. Jarvis; his extant diary begins on January 1, 1868, at Truro, NS, Archives of Manitoba; Sandford Fleming, *The Intercolonial* (Montreal: Dawson Brothers, 1876).

12 Nova Scotia Railway Heritage Society, "Wentworth...then and now" (Elmsdale, NS, n.d.); Jarvis, diary for 1868, Archives of Manitoba.

13 Jarvis diaries, Archives of Manitoba.

14 Fleming, *Canadian Pacific Railway: Report of Progress*; Fleming, *Report on Surveys and Preliminary Operations.*

15 Major C.F. Hanington, Reminiscences 1871–1928, Royal BC Museum, BC Archives, Victoria, BC, E/E/H19 (unpublished).

16 Jarvis, diary entry for June 7, 1875, Archives of Manitoba.

17 C.F. Hanington, journal, May 4, 1876; see Chapter 2.

18 Jarvis, diary entries for May and June 1873, Archives of Manitoba.

19 *Henderson's Directory of the City of Winnipeg and the Incorporated Towns of Manitoba* (Winnipeg: James Weidman at the Inter-Ocean Office, Selkirk, 1880), 193.

20 Alexander Begg and Walter R. Nursey, *Ten Years in Winnipeg* (Winnipeg: Times Printing and Publishing House, 1879), 158.

21 James Elder Steen, *Winnipeg Manitoba and Her Industries* (Winnipeg: Steen and Boyce, 1882), 42.

22 *Times* (Winnipeg, MB), May 23, 1880, http://www.wanlessweb.org/TNG/showmedia.php?mediaID=923.

23 Province of Ontario, Sessional Papers, Vol. 16, Part 7, 1884, p. 47.

24 Jarvis, diary for 1875, Archives of Manitoba; Robert Machray, *Life of Robert Machray* (Toronto: Macmillan Company, 1909), 277.

25 Begg and Nursey, *Ten Years in Winnipeg*; Ruben C. Bellan, "Rails across the Red – Selkirk or Winnipeg" (Manitoba Historical Society Transactions, Series 3, 1961–62 Season).

26 Martin Kavanagh, *The Assiniboine Basin* (Winnipeg: Manitoba Historical Society, 1967), 214.

27 *Henderson's Directory*, 189–191.

28 ICE, Obituary for E.W. Jarvis; Manitoba Historical Society, Memorable

Manitobans: Edward Worrell Jarvis (c. 1846–1894), http://www.mhs.mb.ca/docs/people/jarvis_ew.shtml.

29 Virtual Museum of Canada, Back to Batoche, http://www.museevirtuel-virtualmuseum.ca/sgc-cms/expositions-exhibitions/batoche/html/resources/proof_order_of_battle.php; Mary Beacock Fryer, *Battlefields of Canada* (Toronto: Dundurn Press, 1986).

30 Dominion of Canada, *The Queen vs Louis Riel, Accused and Convicted of the Crime of High Treason* (Ottawa: Queen's Printer, 1886), 90–91.

31 University of Saskatchewan Library, The Batoche Diary, http://library.usask.ca/northwest/diary/riel-dry.htm.

32 Email correspondence with Mary Macdonald, granddaughter of the discoverer, George Waight, on December 22, 2012, and December 24, 2013.

33 Jarvis probate record, Archives of Manitoba.

34 University of Saskatchewan Library, The Batoche Diary.

35 ICE, Obituary for E.W. Jarvis; Canada, Department of Militia and Defence, *Report upon the Suppression of the Rebellion in the North-West Territories and Matters in Connection Therewith, in 1885* (Ottawa, 1886), 34.

36 Dominion of Canada, Parliamentary Sessional Papers, Vol. 24, Issue 15, Ottawa, 1891, p. 22.

37 Dominion of Canada, *Report of the Commissioner of the North-West Mounted Police Force 1886*, Appendix J (Ottawa, 1887), 61.

38 E.W. Jarvis North West Mounted Police personnel files, Library and Archives Canada, http://www.bac-lac.gc.ca/eng/discover/nwmp-personnel-records/Pages/item.aspx?IdNumber=37565&.

39 Dominion of Canada, Parliamentary Sessional Papers, Vol. 24, Issue 15, Ottawa, 1891.

40 Dominion of Canada, Parliamentary Sessional Papers, Vol. 9, Issue 15, Ottawa, 1893.

41 His will was dated February 23, 1893, in Calgary.

42 The City of Calgary, Historical Information, http://www.calgary.ca/CA/city-clerks/Pages/Corporate-records/Archives/Historical-information/Historical-Information.aspx; Calgary, *Wikipedia*, http://en.wikipedia.org/wiki/Calgary.

43 Bow Valley Ranche at Fish Creek Provincial Park, Pack of Western Wolves, http://www.bowvalleyranche.com/pack-wolves.html; Glenbow Museum, photo collection of Western Wolves, http://ww2.glenbow.org/search/archivesPhotosResults.aspx.

44 E.W. Jarvis's personnel files reveal that Commissioner Herchmer travelled to Calgary when he heard of Jarvis's illness; various officers travelled to his funeral; his Calgary press death notice referred to him as "the popular officer"; prayers were offered in the Anglican church for the recovery of Jarvis, who was Roman Catholic.

45 E.W. Jarvis, obituary, *Winnipeg Free Press*, November 27, 1894; E.W. Jarvis North West Mounted Police personnel files.

46 Ibid.

47 Ibid.

48 RCMP Graves, http://www.rcmpgraves.com.

49 Jarvis probate record, Archives of Manitoba; Sam McBride, "Cousin E.W. Jarvis Had a Dramatic Life of Accomplishments and Adventure in the Canadian Frontier," *thebravestCanadian* (blog), December 12, 2014, https://thebravestcanadian.wordpress.com/2014/12/12/mystery-jarvis-gray-cousin-had-a-dramatic-life-of-accomplishments-in-the-canadian-frontier/.

50 "Happy day, moving some furniture into the house – and concerning other things." Jarvis, diary entry for August 11, 1875, Archives of Manitoba.

51 Jarvis, diary entry for August 24, 1875, Archives of Manitoba; Machray, *Life of Robert Machray*, 277.

52 Jarvis, diary entries for September 2 and 12, 1875, Archives of Manitoba.

53 Canada, Library and Archives Canada, 1881 Census of Canada for Manitoba, District 183 Selkirk, p. 34.

54 Manitoba Historical Society, Memorable Manitobans; Manitoba Historical Society, Red River Bridges, http://www.mhs.mb.ca/docs/sites/redriverbridges.shtml.

55 See BC Geographical Names, http://geobc.gov.bc.ca/base-mapping/atlas/bcnames/.

56 Alberta Environment and Sustainable Resource Development, Alberta Parks, http://www.albertaparks.ca; Hinton: Gateway to the Rockies, http://www.hinton.ca/DocumentCenter/View/1719.

57 Begg and Nursey, *Ten Years in Winnipeg*, 192; Old Mersey Times, Liverpool Passenger Lists 1874, http://www.old-merseytimes.co.uk/passengerlists1874.html.

58 Hanington, journal, May 4, 1876; see Chapter 2.

59 Hanington, Reminiscences 1871–1928.

60 E.W. Jarvis, baptismal record, PARO Collections Database, Public

Archives of Prince Edward Island, http://www.gov.pe.ca/archives/parosearch/vital/individual-vital-information/recordId/152452/eventType/1.

61 *Henderson's Directory*, 191.

62 Library and Archives Canada, R4835 – 0-X-E, http://data2.archives.ca/pdf/pdf001/p000001707.pdf; his brother's children are mentioned in his will (Jarvis probate record, Archives of Manitoba).

63 Hanington, Reminiscences 1871–1928.

64 "Featherston/Hanington Family Lineage" (in possession of Hanington family, n.d.).

65 "Journal of Mr. C.F. Hanington from Quesnelle through the Rocky Mountains, during the Winter of 1874–5." It forms Chapter 2 of this book.

66 Hanington, Reminiscences 1871–1928.

67 Berton, *The National Dream*, 415.

68 "Featherston/Hanington Family Lineage."

69 "Featherston/Hanington Family Lineage"; the family history lists his birth year as 1847, but a very precise age on his burial records suggests 1848, as does the 1851 census. His obituary also lists April 14, 1848. A.L. Herne, "Blazing Canada's Trail of Steel [obituary of C.F. Hanington]," *Vancouver Province*, February 8, 1931, p. 9. See also Daniel Hanington, *Wikipedia*, http://en.wikipedia.org/wiki/Daniel_Hanington, and Daniel Lionel Hanington, *Wikipedia*, http://en.wikipedia.org/wiki/Daniel_Lionel_Hanington.

70 Hanington, Reminiscences 1871–1928.

71 Hanington, Reminiscences 1871–1928; *Daily Telegraph* (St. John), June 16, 1871, http://archives.gnb.ca/Search/NewspaperVitalStats/Details.aspx?culture=en-CA&guid=7C1A54FB-44B5-4D49-8801-ADDB0873ED95.

72 Hanington, Reminiscences 1871–1928; Fleming, *Canadian Pacific Railway: Report of Progress*; Fleming, *Report on Surveys and Preliminary Operations*.

73 Hanington, Reminiscences 1871–1928; Hanington, journal, May 4, 1876; see Chapter 2 of this book.

74 *Chignecto Post* (Sackville, NB), November 8, 1877, http://archives.gnb.ca/Search/NewspaperVitalStats/Details.aspx?culture=en-CA&guid=d19c134d-ec0a-4fa4-b346-5ced0d421db6&r=1&ni=126951.

75 Hanington, Reminiscences 1871–1928.

76 "Featherston/Hanington Family Lineage"; 1891 census, http://data2.collectionscanada.ca/1901/z/z001/jpg/z000013423.jpg.

77 Hanington, Reminiscences 1871–1928.

78 Dominion of Canada, Parliamentary Sessional Papers, Vol. 48, Issue 25, Victoria, 1885, p. 11.

79 Hanington, Reminiscences 1871–1928; *Chignecto Post* (Sackville, NB), September 4, 1884, http://archives.gnb.ca/Search/NewspaperVitalStats/Details.aspx?culture=en-CA&guid=28962dfb-16af-460a-a7bf-2a0d8cc52b68&r=1&ni=126948.

80 Hanington, Reminiscences 1871–1928; "Featherston/Hanington Family Lineage"; *The Dominion Annual Register* (Toronto: Hunter Rose and Co., 1885), 392. The site of the mill that Hanington purchased is currently occupied by the Shediac Visitor Centre and the "World's Largest Lobster," a sculpture that forms a tourist attraction.

81 Hanington, Reminiscences 1871–1928.

82 *Times* (Moncton, NB), May 6, 1889, http://archives.gnb.ca/Search/NewspaperVitalStats/Details.aspx?culture=en-CA&guid=84cbe6dd-4054-4b65-91c9-5d842da3580e&r=1&ni=126948.

83 Ibid.

84 New Brunswick Archives RS581: Index to Justice of the Peace Appointment Register, 1863–1963.

85 Hanington, Reminiscences 1871–1928.

86 Ibid.

87 Hanington, Reminiscences 1871–1928; *Prospector* (Fort Steele), November 18, 1899, http://historicalnewspapers.library.ubc.ca/view/collection/prospector/date/1899-11-18#1.

88 1901 census, http://data2.collectionscanada.ca/1901/z/z001/jpg/z000013423.jpg.

89 Hanington, Reminiscences 1871–1928.

90 Herne, "Blazing Canada's Trail of Steel."

91 Hanington, Reminiscences 1871–1928.

92 Ibid.

93 1911 census, http://data2.collectionscanada.gc.ca/1911/jpg/e002009527.jpg.

94 Hanington, Reminiscences 1871–1928.

95 Ibid.

96 Hanington, Reminiscences 1871–1928; O.M. Meehan, "The Hydrographic Survey of Canada from the First World War to the Commencement of the Canadian Hydrographic Service, 1915–1927," *The Northern Mariner/Le marin du nord* 14, no. 1 (January 2004): 109, http://www.cnrs-scrn.org/

northern_mariner/vol14/tnm_14_1_105-158.pdf; Friends of Hydrography, https://www.google.ca/#q=canfoh+%2B+hanington.

97 Hanington, Reminiscences 1871–1928; Attestation Papers, Canadian Expeditionary Force, http://www.bac-lac.gc.ca/eng/discover/military-heritage/first-world-war/first-world-war-1914-1918-cef/Pages/item.aspx?IdNumber=444073.

98 Herne, "Blazing Canada's Trail of Steel."

99 Attestation Papers, Canadian Expeditionary Force.

100 Hanington, Reminiscences 1871–1928.

101 "Featherston/Hanington Family Lineage."

102 Herne, "Blazing Canada's Trail of Steel."

103 Hanington, Reminiscences 1871–1928; Friends of Hydrography.

104 Herne, "Blazing Canada's Trail of Steel."

105 1911 census; All Results for Charles Francis Hanington, Ancestry, http://search.ancestry.co.uk/cgi-bin/sse.dll?gl=allgs&gss=sfs28_ms_f-2_s&new=1&rank=1&msT=1&gsfn=charles%20francis&gsln=hanington&mswpn_ftp=Canada&mswpn=3243&mswpn_PInfo=3-%7C0%7C1652393%7C0%7C3243%7C0%7C0%7C0%7C0%7C0%7C&msbdy_x=1&msbdp=5&MSAV=1&msbdy=1848&cpxt=1&cp=3&catbucket=rstp&uidh=5q4.

106 Burial records were obtained from ancestry.ca at http://trees.ancestry.co.uk/tree/22255417/person/1353121127; All Results for Charles Francis Hanington, *Ancestry*.

107 See BC Geographical Names, http://geobc.gov.bc.ca/base-mapping/atlas/bcnames/.

108 See https://ab-hinton.civicplus.com/DocumentCenter/View/1719.

109 Association of BC Land Surveyors, *The L.S. Group: British Columbia's First Land Surveyors* (Sidney, BC: Association of BC Land Surveyors, 2007).

110 Jarvis, diary for 1873, Archives of Manitoba; Hanington, Reminiscences 1871–1928.

111 Hanington, Reminiscences 1871–1928.

112 Ibid.

113 Ibid.

114 Fleming, *Report on Surveys and Preliminary Operations.*

115 See Chapter 2.

116 Ibid.

117 See Chapter 1.

118 Edmonton Historical Board, "Honouring Heritage Achievement."

119 Our sources for our biography of Sandford Fleming include Sandford Fleming, *Wikipedia*, http://en.wikipedia.org/wiki/Sandford_Fleming; Lawrence J. Burpee, *Ryerson Canadian History Readers: Sir Sandford Fleming* (Toronto: The Ryerson Press, 1930); Berton, *The National Dream*.

120 Burpee, *Ryerson Canadian History Readers*, 1–2.

121 Sandford Fleming, "Appendix A: Observations and practical suggestions on the subject of a railway through British North America," in *Memorandum on the Canadian Pacific Railway*, submitted to the government of the province of Canada in the year 1863 (Ottawa, 1874).

APPENDIX A
THE SMOKY RIVER CACHE

1 See Chapter 1.

2 See Chapter 1.

3 See Chapter 1.

4 Jarvis, diary for 1875, Archives of Manitoba. This list is not reproduced in Chapter 3.

5 Jarvis, diary entry for March 6, 1875, Archives of Manitoba; see Chapter 3.

6 Jarvis, diary for 1875, Archives of Manitoba; it is not reproduced in Chapter 3 as it postdates the expedition, after Jarvis had resigned from the survey.

APPENDIX B
THE JARVIS DIARIES

1 Olive Knox, *The Young Surveyor* (Toronto: Ryerson Press, 1956).

APPENDIX D
EXCERPTS FROM OTHER WRITERS

1 J.H.E. Secretan, *Canada's Great Highway: From the First Stake to the Last Spike* (Ottawa: Thorburn & Abbott, 1924), 69.

2 Charles Frederick Carter, *When Railroads Were New* (New York: Henry Holt and Company, 1909), 295–296.

3 Knox, *The Young Surveyor.*

4 Ibid., 163–164.

5 Excerpted from *The National Dream: The Great Railway, 1871–1881* by Pierre Berton. Copyright © 1970 Pierre Berton Enterprises Ltd. Reprinted by permission of Anchor Canada/Doubleday Canada, a division of Random House of Canada Limited, a Penguin Random House Company.

6 Barry Cotton, introduction to Yvonne Klan's "We Are Travelling through an Unknown Country," *British Columbia Historical News* 35, no. 1 (Winter 2001/2002): 8.

7 Yvonne Klan, "We Are Travelling through an Unknown Country," *British Columbia Historical News* 35, no. 1 (Winter 2001/2002): 8–13.

8 James McGregor, "Edward William [*sic*] Jarvis: Prince George to Fort Edmonton in 1875," *Alberta Historical Review* 6, no. 1 (Winter 1958): 1–9.

9 From *Great Railroad Tunnels of North America* © 2011 William Lowell Putnam by permission of McFarland & Company, Inc., Box 611, Jefferson NC 28640.

APPENDIX E
MAL DE RAQUETTE

1 See Chapter 2.

2 See Chapter 1.

3 "Mal Due Raquette – snowshoe sickness," *Nor'west Scribe* (blog), December 17, 2010, http://norwestscribe-voyager.blogspot.ca/2010/12/mal-due-raquette-snowshoe-sickness.html.

APPENDIX F
GEOGRAPHIC NAMES

1 See Chapter 1.

2 Frederick Vreeland, "Early Visits to Mount Sir Alexander," *American Alpine Journal*, (1930): 114–119.

3 S. Prescott Fay, "Mt Alexander," *Canadian Alpine Journal* VI (1914–1915): 121.

4 Mount Ida (Crete), *Wikipedia*, http://en.wikipedia.org/wiki/Mount_Ida_%28Crete%29.

5 Alice Dunn and Fred Dunn, "First Ascent of Mount Ida," *Canadian Alpine Journal* XXXVIII (1955): 33.

6 Knox, *The Young Surveyor*, 127.

7 Jarvis Pass, GeoBC, http://apps.gov.bc.ca/pub/bcgnws/names/2929.html.

8 Mount Jarvis, GeoBC, http://apps.gov.bc.ca/pub/bcgnws/names/10059.html.

9 Mount St. David, GeoBC, http://apps.gov.bc.ca/pub/bcgnws/names/13187.html.

10 See Charles Helm and Mike Murtha, eds., *The Forgotten Explorer: Samuel Prescott Fay's 1914 Expedition to the Northern Rockies* (Victoria: Rocky Mountain Books, 2009).

11 Whyte Museum of the Canadian Rockies, Fred Brewster fonds, M53/V86.

APPENDIX G
GEOGRAPHICAL FEATURES

1 For road conditions, check first with the provincial forestry office in Prince George, https://www.for.gov.bc.ca/dpg/; Herrick Falls co-ordinates are 54° 16' 24" north, 121° 06' 13" west.

2 Mary L. Jobe, "A Winter Journey to Mt Sir Alexander and the Wapiti," *Canadian Alpine Journal* IX (1918): 63–65.

3 Mary L. Jobe, "Mt Alexander Mackenzie," *Canadian Alpine Journal* VII (1916): 62.

4 Dunn and Dunn, "First Ascent of Mount Ida," 33.

5 Helm and Murtha, *The Forgotten Explorer*, 48–49.

6 Ibid., 50.

7 Ibid.

8 H.F. Lambart, "The Canadian Rockies from Yellowhead Pass North to Jarvis Pass," *Canadian Alpine Journal* XVIII (1929): 63.

9 A.O. Wheeler and H.F. Lambart, "Mountain Reconnaissance by Airplane," *Canadian Alpine Journal* XIII (1923): 112–119.

10 Theodore J. Holsten and Susan C. Reneau, eds., *From the Peace to the Fraser: Newly Discovered North American Hunting and Exploration Journals by Prentiss N. Gray* (Missoula, MT: Boone and Crockett Club, 1994), 319.

SELECTED BIBLIOGRAPHY

Anderson, A.C. *Upper Fraser River: maps and notes submitted to Marcus Smith in support of a route for the Canadian Pacific Railway.* 1874. BC Archives.

Andrews, Gerry. *Metis Outpost: memoirs of the first schoolmaster at the Metis settlement of Kelly Lake, BC, 1923–1925.* Victoria, BC, 1985.

Association of BC Land Surveyors. Cumulative Nominal Roll of Professional Land Surveyors of the Province of British Columbia. http://www.abcls.ca/wp-content/uploads/.../Nominal_Roll_2012-08-25.pdf.

Association of BC Land Surveyors. *The L.S. Group: British Columbia's First Land Surveyors.* Sidney, BC: Association of BC Land Surveyors, 2007.

Begg, Alexander, and Walter R. Nursey. *Ten Years in Winnipeg.* Winnipeg: Times Printing and Publishing House, 1879.

Bellan, Ruben C. "Rails across the Red – Selkirk or Winnipeg." Manitoba Historical Society Transactions, Series 3, 1961–62 Season.

Berton, Pierre. *The National Dream: The Great Railway, 1871–1881.* Toronto: McClelland and Stewart Ltd., 1970.

Brymner, Douglas. Report on Canadian Archives 1887. Parliament of Canada, Sessional Papers, Volume 5. Ottawa, 1888.

Bumsted, J.M. "The Household and Family of Edward Jarvis, 1828–1852." *Island Magazine*, no. 14 (Fall–Winter 1983): 22–28.

Bumsted, J.M., and H.T. Holman. "JARVIS, EDWARD JAMES." In *Dictionary of Canadian Biography*, vol. 8, University of Toronto/Université Laval, 2003–. http://www.biographi.ca/en/bio/jarvis_edward_james_8E.html.

Burpee, Lawrence J. *Ryerson Canadian History Readers: Sir Sandford Fleming,* Toronto: The Ryerson Press, 1930.

Canada, Department of Militia and Defence. *Report upon the Suppression of the Rebellion in the North-West Territories and Matters in Connection Therewith, in 1885.* Ottawa, 1886.

Carter, Charles Frederick. *When Railroads Were New.* New York: Henry Holt and Company, 1909.

Census of Canada for Manitoba, 1881. Library and Archives Canada. http://www.bac-lac.gc.ca/eng/Census/1881/Pages/mb.aspx.

Cotton, Barry. Introduction to "We Are Travelling through an Unknown Country," by Yvonne Klan. *British Columbia Historical News* 35, no. 1 (Winter 2001/2002): 8.

Dawson, G.M. "Report on an exploration from Port Simpson on the Pacific coast to Edmonton on the Saskatchewan, embracing a portion of the northern part of British Columbia and the Peace River country." In *Geological and Natural History Survey of Canada: Report of Progress for 1879–80*, 1–95. Montreal: Dawson Bros., 1881.

The Dominion Annual Register. Toronto: Hunter Rose and Co., 1885.

Dominion of Canada. Parliamentary Sessional Papers, Vol. 48, Issue 25. Victoria, 1885.

Dominion of Canada. Parliamentary Sessional Papers, Vol. 24, Issue 15. Ottawa, 1891.

Dominion of Canada. Parliamentary Sessional Papers, Vol. 9, Issue 15. Ottawa, 1893.

Dominion of Canada. *The Queen vs Louis Riel, Accused and Convicted of the Crime of High Treason*. Ottawa: Queen's Printer, 1886.

Dominion of Canada. *Report of the Canadian Pacific Railway Royal Commission*. Volume 1 Evidence. Ottawa, 1882.

Dominion of Canada. *Report of the Commissioner of the North-West Mounted Police Force 1886*. Ottawa, 1887.

Dunn, Alice, and Fred Dunn. "First Ascent of Mount Ida." *Canadian Alpine Journal* XXXVIII (1955).

E.W. Jarvis North West Mounted Police personnel files. Library and Archives Canada. http://www.bac-lac.gc.ca/eng/discover/nwmp-personnel-records/Pages/item.aspx?IdNumber=37565&.

Edmonton Historical Board. "Honouring Heritage Achievement." *Building Heritage* 2, no. 2 (December 1998). http://www.edmonton.ca/city_government/documents/buildingheritagenews_vol_2_no_2_dec_1998.pdf.

Elliott, Gordon R. *Quesnel, Commercial Centre of the Cariboo Gold Rush*. Quesnel, BC: Cariboo Historical Society, 1958.

Fay, S. Prescott. "Mt Alexander." *Canadian Alpine Journal* VI (1914–1915): 170–177.

Fleming, Sandford. "Appendix A: Observations and practical suggestions on the subject of a railway through British North America." In *Memorandum on the Canadian Pacific Railway*, 21–49. Submitted to the government of the province of Canada in the year 1863. Ottawa, 1874.

Fleming, Sandford. *Canadian Pacific Railway: Report of Progress on the Explorations and Surveys up to January 1874*. Ottawa: MacLean, Roger & Co., 1874.

Fleming, Sandford. "Expeditions to the Pacific." In *Proceedings and Transactions of the Royal Society of Canada for the Year 1889*, Vol. 7, Section 2, Part 4, 89–141. Montreal: Dawson Brothers, 1890.

Fleming, Sandford. *The Intercolonial*. Montreal: Dawson Brothers, 1876.

Fleming, Sandford. *Report on Surveys and Preliminary Operations of the Canadian Pacific Railway up to January 1877*. Ottawa, 1877. Library and Archives Canada.

Friends of Hydrography. https://www.google.ca/#q=canfoh+%2B+hanington.

Friesen, Victor Carl. *Where the River Runs – Stories of the Saskatchewan and the People Drawn to Its Shores*. Calgary: Fifth House Press, 2001.

Fryer, Mary Beacock. *Battlefields of Canada*. Toronto: Dundurn Press, 1986.

Hanington, C.F. Correspondence 1928–1929. BC Archives.

Hanington, C.F. "Journal of Mr. C.F. Hanington from Quesnelle through the Rocky Mountains, during the Winter of 1874–5." Note C in Report on Canadian Archives 1887. Dominion of Canada, Sessional Papers, Vol. 5, Second Session of Sixth Parliament. Ottawa, 1888.

Hanington, C.F. Reminiscences 1871–1928. BC Archives.

Hayes, Derek. *British Columbia: A New Historical Atlas*. Vancouver: Douglas & McIntyre, 2012.

Helm, Charles, and Mike Murtha, eds. *The Forgotten Explorer: Samuel Prescott Fay's 1914 Expedition to the Northern Rockies*. Victoria: Rocky Mountain Books, 2009.

Henderson's Directory of the City of Winnipeg and the Incorporated Towns of Manitoba. Winnipeg: James Weidman at the Inter-Ocean Office, Selkirk, 1880.

Historical Atlas of Canada Online Learning Project. http://www.historicalatlas.ca/website/hacolp/national_perspectives/exploration/UNIT_08_5/index.htm.

Holsten, Theodore J., and Susan C. Reneau, eds. *From the Peace to the Fraser: Newly Discovered North American Hunting and Exploration Journals by Prentiss N. Gray*. Missoula, MT: Boone and Crockett Club, 1994.

Index to Justice of the Peace Appointment Register, 1863–1963. Provincial Archives of New Brunswick. http://archives.gnb.ca/Search/RS581/Details.aspx?culture=en-CA&record=4487.

Institution of Civil Engineers (ICE). Election as associate member: Minutes of Proceedings, Vol. XXXVIII, Part II. Session 1873–1874.

Institution of Civil Engineers (ICE). Obituary for E.W. Jarvis: Minutes of the Proceedings, Vol. 124, Issue 1896. January 1896.

Jarvis, E.W. Baptism Record. PARO Collections Database. Public Archives of Prince Edward Island. http://www.gov.pe.ca/archives/parosearch/vital/individual-vital-information/recordId/152452/eventType/1.

Jarvis, E.W. Diaries. Archives of Manitoba.

Jarvis, E.W. "Narrative of the Exploration from Fort George, across the Rocky Mountains, by Smoky River Pass to Manitoba, Referred To in the Preceding Report." In *Report on Surveys and Preliminary Operations of the Canadian Pacific Railway up to January 1877*, Appendix H, 148–161. Ottawa, 1877.

Jarvis, E.W. Probate Record. Archives of Manitoba. http://www.gov.mb.ca/chc/archives/probate/wpg_estate.html#wpg1870.

Jarvis, E.W. "Report on Exploration across the Rocky Mountains by Smoky River Pass." In *Report on Surveys and Preliminary Operations of the Canadian Pacific Railway up to January 1877*, Appendix H, 145–147. Ottawa, 1877.

Jarvis, George A., George Murray Jarvis, and William Jarvis Wetmore. *The Jarvis Family*. Hartford, CT: Case, Lockwood and Brainard, 1879.

Jarvis, Henry Fitzgerald fonds (textual record). Library and Archives Canada. http://data2.archives.ca/pdf/pdf001/p000001707.pdf.

Jobe, Mary L. "Mt Alexander Mackenzie." *Canadian Alpine Journal* VII (1916).

Jobe, Mary L. "A Winter Journey to Mt Sir Alexander and the Wapiti." *Canadian Alpine Journal* IX (1918): 63–65.

Kavanagh, Martin. *The Assiniboine Basin*. Winnipeg: Manitoba Historical Society, 1967.

Klan, Yvonne. "We Are Travelling through an Unknown Country." *British Columbia Historical News* 35, no. 1 (Winter 2001/2002): 8–13.

Knox, Olive. *The Young Surveyor*. Toronto: Ryerson Press, 1956.

Lambart, H.F. "The Canadian Rockies from Yellowhead Pass North to Jarvis Pass." *Canadian Alpine Journal* XVIII (1929).

Machray, Robert. *Life of Robert Machray*. Toronto: Macmillan Company, 1909.

Manitoba Historical Society. Memorable Manitobans: Edward Worrell Jarvis (c. 1846–1894). http://www.mhs.mb.ca/docs/people/jarvis_ew.shtml.

Manitoba Historical Society. Red River Bridges. http://www.mhs.mb.ca/docs/sites/redriverbridges.shtml.

McGregor, James. "Edward William [*sic*] Jarvis: Prince George to Fort Edmonton in 1875." *Alberta Historical Review* 6, no. 1 (Winter 1958): 1–9.

Meehan, O.M. "The Hydrographic Survey of Canada from the First World War to the Commencement of the Canadian Hydrographic Service, 1915–1927." *The Northern Mariner/Le marin du nord* 14, no. 1 (January 2004): 105–157. http://www.cnrs-scrn.org/northern_mariner/vol14/tnm_14_1_105-158.pdf.

Moberly, Henry John. *When Fur Was King*. London and Toronto: J.M. Dent, 1929.

Moberly, Walter. *The Rocks and Rivers of British Columbia*. London: H. Blacklock and Co., 1885.

Nova Scotia Railway Heritage Society. "Wentworth...then and now." Elmsdale, NS, n.d.

Old Mersey Times. Liverpool Passenger Lists 1874. http://www.old-merseytimes.co.uk/passengerlists1874.html.

Province of Ontario. Sessional Papers, Vol. 16, Part 7. 1884.

Putnam, William Lowell. *Great Railroad Tunnels of North America*. Jefferson, NC: McFarland & Company, 2011.

Rich, E.E., ed. *Colin Robertson's correspondence Book, Sept. 1817 to Sept. 1822*. Toronto: The Champlain Society, 1939.

Secretan, J.H.E. *Canada's Great Highway: From the First Stake to the Last Spike*. Ottawa: Thorburn & Abbott, 1924.

Selwyn, Alfred R.C., "Report on Exploration in British Columbia." In *Geological Survey of Canada: Report of Progress 1875–1876*, 28–86. Montreal: Dawson Brothers, 1877.

Smith, Marcus. Correspondence. BC Archives.

Smith, Marcus. "Special Report on Passes through the Cascade and Rocky Mountain Chains." Appendix K in *Canadian Pacific Railway: Report of Progress on the Explorations and Surveys up to January 1874*, 216–217. Ottawa: MacLean, Roger & Co., 1874.

Spiral Road. http://www.spiralroad.com/smoky-river/.

Steele, Sam. *Forty Years in Canada: Reminiscences of the Great Northwest with Some Account of His Service in South Africa*. Toronto: McClelland, Goodchild, & Stewart, 1915.

Steen, James Elder. *Winnipeg Manitoba and Her Industries*. Winnipeg: Steen and Boyce, 1882.

Telfer, W.P. *Humboldt on the Carlton Trail*. Saskatoon: Modern Press, 1975.

University of Saskatchewan Library. The Batoche Diary. http://library.usask.ca/northwest/diary/riel-dry.htm.

Virtual Museum of Canada. Back to Batoche. http://www.museevirtuel-virtualmuseum.ca/sgc-cms/expositions-exhibitions/batoche/html/resources/proof_order_of_battle.php.

Vreeland, Frederick. "Early Visits to Mount Sir Alexander." *American Alpine Journal* (1930): 114–119.

Wheeler, A.O., and H.F. Lambart. "Mountain Reconnaissance by Airplane." *Canadian Alpine Journal* XIII (1923): 112–119.

INDEX

Note: Locators in bold italics indicate an illustration; locators in the form 234n24 indicate an endnote (in this case, note 24 on page 234; 233nn17–18 means notes 17 and 18 on page 233)

Abbreviations: CPR is Canadian Pacific Railway; HBC is Hudson's Bay Company; J-H Expedition is Jarvis–Hanington Winter Expedition; NWMP is North West Mounted Police